THE
NEW PHOENIX
LIBRARY

*

Fiery Particles

Fiery Particles

C. E. MONTAGUE

THE NEW PHOENIX

Published by

CHATTO & WINDUS

LONDON

★

CLARKE, IRWIN & CO. LTD

TORONTO

First published 1923
Reprinted 1923 (twice), 1925, 1927,
1928, 1930, 1936
First published in this Library 1951

To

THE MEMORY OF

W. T. A.   L. P. S.

S. W. H.

UNLIKE ONE ANOTHER IN EVERYTHING
BUT THAT THEY WERE NOT RULED
BY FEAR OR DESIRE
AND YOU COULD BELIEVE
WHAT THEY SAID

A NOTE

NOW that I see them set out in a row, these yarns seem to be all about a set of wild bodies that want to be up and doing something, as often foolish as not : everywhere somebody much taken up with a lance that he has——shining or shabby, he wants to put it in rest; he rides out on some good or queer quest, in a great state of absorption and hope, pricking a hobby-horse bred, and imperfectly broken in, by himself. I blush to find my creatures so forward; they step up to life, they speak to her first and offer to print their own whims on such talk as may pass between them and her before she consign, them to dust.

I fear these ardent cranks are a little out of fashion. They ought numbly to suffer the business or game of living, not pull it about nor try to give it new twists, each to his own wayward liking. Ours is the day of the hero who slips through life; voluble, yes; but passive, a drifter, pleading that he is the fault of everyone else and declining all of life that is declinable. Still, what is a fellow to do ? If, of all the men you have known, none will come back to your mind except arrant lovers of living, mighty hunters of lions or shadows, rapt amateurs of shady adventure or profitless zeal, how can you steep them in languor enough to bring them up to the mark ? Better let them go, and take their chance, as the fiery particles that they were in the flesh.

CONTENTS

ANOTHER TEMPLE GONE

I

THEY say that there may be a speck of quiet lodged at the central point of a cyclone. Round it everything goes whirling. It alone sits at its ease, as still as the end of an axle that lets the wheel, all about it, whirl any wild way it likes.

That was the way at Gartumna in those distant years when the " land war " was blowing great guns all over the rest of the County Clare. Gartumna lay just at the midst of that tempest. But not a leaf stirred in the place. You paid your rent if you could ; for the coat that the old colonel had on his back—and he never out of the township—was that worn you'd be sorry. Suppose you hadn't the cash, still you were not " put out of it." All that you'd have to suffer was that good man buzzing about your holding, wanting to help ; he would be all in a fidget trying to call to mind the way that some heathen Dane, that he had known when a boy, used to bedevil salt butter back into fresh—that, or how Montenegrins would fatten a pig on any wisp of old trash that would come blowing down the high road. A kind man, though he never got quit of the queer dream he had that he knew how to farm.

Another practising Christian we had was Father O'Reilly. None of the sort that would charge you half the girl's fortune before they'd let the young people set foot in the church. And, when it was done, he'd come to the party and sing the best song of anyone there. However, at practical goodness Tom Farrell left the entire field at the post. Tom had good means : a farm in fee-simple—

the land, he would often tell us, the finest in Ireland,
" every pitaty the weight of the world if you'd take it up
in your hand "; turf coming all but in at the door to be
cut; besides, the full of a creel of fish in no more than the
time you'd take dropping a fly on the stream: the keeper
had married Tom's sister. People would say " Ach, the
match Tom would be for a girl ! " and gossips liked count-
ing the " terrible sum " that he might leave when he'd die
if only he knew how to set any sort of value on money.
But this he did not. The widow Burke, who knew
more about life than a body might think, said Tom would
never be high in the world because no one could come and
ask for a thing but he'd give it them. Then, as she warmed
to the grateful labour of letting you know what was what,
the widow might add: " I question will Tom ever make
a threepenny piece, or a penny itself, out of that old con-
struction he has away there in the bog."

At these words a hearer would give a slight start and
glance cannily round, knowing that it would be no sort
of manners to give a decent body like Sergeant Maguire
the botheration and torment of hearing the like of that
said out aloud. But the sergeant would never be there.
For he too had his fine social instincts. He would be half
a mile off, intent on his duty, commanding the two decent
lads that were smoking their pipes, one on each of his
flanks, in the tin police hut away down the road. Gartumna
did not doubt that this tactful officer knew more than he
ever let on. A man of his parts must surely have seen, if
not smelt, that no unclean or common whisky, out of a
shop, had emitted the mellow sunshine transfiguring recent

2

christenings and wakes. But who so coarse as to bring a functionary so right-minded up against the brute choice between falling openly short in professional zeal and wounding the gentle bosom of Gartumna's peace?

And yet the widow's sonorous soprano, or somebody else's, may have been raised once too often on this precarious theme. For, on one of the warmest June mornings that ever came out of the sky, Sergeant Maguire paraded his whole army of two, in line, on a front of one mile, with himself as centre file and file of direction, and marched out in this extremely open order into the fawn-coloured wilderness of the bog. "You'll understand, the two of yous," he had said to his right flank, Constable Boam, and to Constable Duffy, his left, "that this is a sweeping or dhragging movement that we're making."

The sun was high already—your feverish early starts were no craze of the sergeant's. The air over the bog had tuned up for the day to its loudest and most multitudinous hum and hot click of grasshoppers and bees; all the fawn surface swam in a water-coloured quiver of glare; the coarse, juiceless grass and old roots, leathery and slippery, tripped up the three beaters' feet. Hour by hour the long morning greased and begrimed the three clean-shaven, good-soldier faces that had set out on the quest; noon came blazingly on—its savage vertical pressure seemed to quell and mute with an excess of heat the tropical buzz of all the basking bog life that the morning's sunshine had inspirited; another hour and the bog was swooning, as old poets say, under the embraces of the sun her friend, when a thin column of more intensely quivering air, a

3

hundred yards off to the sergeant's half-left, betrayed some source of an ardour still more fiery than the sun's. Just for the next five or ten minutes, no more, the sergeant had some good stalking. Then it was all over. The hunting was done : nothing left but to whistle in his flank men and go over the haul.

The tub and worm of the illicit still had not been really hidden ; they were just formally screened with a few blocks of turf as though in silent appeal to the delicacy of mankind to accept as adequate this symbolic tribute to the convention of a seemly reticence. Farrell, a little, neatly-made, fine-featured man with a set, contained face, but with all the nervousness of him quivering out into the rest-less tips of his small, pointed fingers, gazed at the three stolid uniformed bulks, so much grosser than he, while they disrobed his beloved machinery of that decent light vesture of turf and rummaged with large, coarse hands among the mysteries of his craft. He wore the Quakerish black suit and the broad and low-crowned soft black hat in which a respectable farmer makes his soul on a Sunday morning. Silent, and seemingly not shamed, nor yet enraged, neither the misdemeanant caught in the act nor the parent incensed by a menace to its one child, he looked on, grave and almost compassionate. So might the high priestess of Vesta have looked when the Gaulish heathen came butting into the shrine and messed about with the poker and tongs of the goddess's eternal flame. How could the poor benighted wretches know the mischief that they might be doing the world ?

Sergeant Maguire, too, may have had his own sense of

our kind's tragic blindness quickened just then—that a
man, a poor passionate man, should so rush upon his own
undoing! " Ach, it's a pity of you, Farrell," he presently
said. " A pity! You with the grand means that you have
of your own! An' you distillin' pocheen! "

" Pocheen! " The little, precise, nervous voice of
Farrell ran up into a treble of melancholy scorn. With an
austere quality in his movements he drew a brown stoneware jar from among some heaped cubes of turf that the
barbarians had not yet disarranged. From another recess
he took a squat tumbler. Into this he poured from the jar
enough to fill a liqueur-glass rather smaller than most.
" Tell me," he bade almost sternly, holding the tumbler
out to Maguire, " d'ye call that pocheen? "

" Ye can take a sup first," was the canny reply.
Maguire had heard how Eastern kings always made cooks
and premiers taste first.

Farrell absorbed the tot, drop by drop. He did not cross
himself first, but there was something about his way of
addressing himself to the draught that would make you
think of a man crossing himself before some devout
exercise, or taking the shoes from off his feet before stepping
on holy ground. As the potion irrigated his soul he seemed
to draw off from the touch of this clamorous world into
some cloistral retreat. From these contemplative shades
he emerged, controlling a sigh, a little time after the
last drop had done its good office. He poured out for
Maguire.

" Well, here's luck," said the sergeant, raising the
glass, " and a light sentence beyond." The good fellow's

tone conveyed what the etiquette of the service would not allow him to say—that in the day of judgment every mitigating circumstance would be freshly remembered.

Up to this his fortieth year Maguire, conversing with the baser liquors of this world and not with philtres of transfiguration, had counted it sin to drink his whisky as if it would burn him. So the whole of the tot was now about to descend his large-bore throat in close order, as charges of shot proceed through the barrel of a gun. But the needful peristaltic action of the gullet had scarcely commenced when certain tidings of great joy were taken in at the palate and forwarded express to an astonished brain. "Mother of God!" the sergeant exclaimed. "What sort of hivven's delight is this you've invented for all souls in glory?"

A sombre satisfaction gleamed out of Farrell's monkish face. Truth was coming into its own, if only too late. The heathen were seeing the light. "It's the stuff," he said, gravely, "that made the old gods of the Greeks and Romans feel sure they were gods."

"Be cripes, they were right," asseverated Maguire. He was imbibing drop by drop now, as the wise poets of all times have done, and not as the topers, the swillers of cocktails, punch and cup, and the like, things only fit to fill up the beasts that perish. Not hoggishness only, but infinite loss would it have seemed to let any one drop go about its good work as a mere jostled atom, lost in a mob of others. If ever the bounty of heaven should raise a bumper crop of Garricks on earth, you would not use them as so many supers, would you?

Farrell, after a short pause to collect his thoughts, was stating another instalment of the facts. "There's a soul and a body," he said, "to everything else, the same as ourselves. Any malt you'll have drunk, to this day, was the body of whisky only—the match of these old lumps of flesh that we're all of us draggin' about till we die. The soul of the stuff's what you've got in your hand."

"It is that," said the sergeant, and chewed the last drop like a lozenge. He now perceived that the use of large, bold, noble figures of speech, like this of Farrell's, was really the only way to express the wonderful thoughts filling up a man's mind when he is at his best. That was the characteristic virtue of Farrell's handiwork. Its merely material parts were, it is true, pleasant enough. They seemed, while you sipped, to be honey, warm sunshine embedded in amber and topaz, the animating essence of lustrous brown velvet, and some solution of all the mellowest varnish that ever ripened for eye or ear the glow of Dutch landscape or Cremona fiddle. No sooner, however, did this potable sum of all the higher physical embodiments of geniality and ardour enter your frame than a major miracle happened in the domain of the spirit : you suddenly saw that the most freely soaring poetry, all wild graces and quick turns and abrupt calls on your wits, was just the most exact, business-like way of treating the urgent practical concerns of mankind.

So the sergeant's receivers were well tuned to take in great truths when Farrell, first measuring out the due dram for Constable Duffy, resumed, "You'll remember the priest that died on us last year ? "

" I do that, rest his soul," said each of the other two Catholics. Constable Boam was only a lad out of London, jumped by some favour into the force. But a good lad.

" Ye'll remember," Farrell continued, " the state he was in, at the end ? Perished with thinness, and he filled with the spirit of God the way you'd see the soul of him shining out through the little worn webbin' of flesh he had on, the match for a flame that's in one of the Chinese lanterns you'd see made of paper. Using up the whole of his body, that's what the soul of him was—convertin' the flesh of it bit by bit into soul till hardly a tittle of body was left to put in the ground. You could lift the whole with a finger."

" Now, aren't ye the gifted man ? " The words seemed to break, of themselves, out of Constable Duffy. Rapt with the view of entire new worlds of thought, and the feel of new powers for tackling them, Duffy gazed open-lipped and wide-eyed at Farrell the giver.

Farrell's face acknowledged, with no touch of wicked pride, this homage to truth. " *Non nobis, Domine.*" Austere, sacerdotal, Farrell inspected the second enraptured proselyte. Then he went on, his eyes well fixed on some object or other far out on the great bog's murmurous waste—the wilfully self-mesmerising stare of the mystic far gone. " The body's the real old curse. Not a thing in the world but it's kept out of being the grand thing it's got the means in it to be if it hadn't a hunk of a body always holding it back. You can't even have all the good there is in a song without some old blether of words would go wrong on your tongue as likely as not. And in Ireland

8

the glory an' wonder that's sent by the will of God to gladden the heart of a man has never got shut till this day of sour old mashes of barley and malt and God alone knows what sort of dish-washin's fit to make a cow vomit, or poisons would blister half of the lining off the inside of an ass."

Constable Duffy was no man of words. But just at this moment he gained his first distinct view of philosophy's fundamental distinction between matter and form; the prospect so ravished his whole being that as he handed in the drained tumbler to Farrell he murmured in a kind of pensive ecstasy, " Hurroosh to your soul ! " and for a long time afterwards was utterly lost in the joys of contemplation.

Constable Boam's reversionary interest in paradise had now matured. While Farrell ministered to Boam, the grapes of the new wine of thought began abruptly to stammer through the lips of the sergeant. " Aye ! Every man has a pack of old trash discommodin' his soul. Pitaties and meal and the like—worked up into flesh on the man. An' the whole of it made of the dirt in the fields, a month or two back ! The way it's a full barrow-load of the land will be walking on every two legs that you'd see shankin' past ! It's what he's come out of. And what he goes back into being. Aye, and what he can't do without having, as long as he lasts. An' yet it's not he. An' yet he must keep a fast hold on it always, or else he'll be dead. An' yet I'll engage he'll have to be fighting it always—it and the sloth it would put on the grand venomous life he has in him. God help us, it's difficult." Along the mazy

9

path that has ever followed in the wake of Socrates the
sergeant's mind slowly tottered, clinging at each turn to
some reminiscence of Farrell's golden words, as a child
makes its first adventurous journey on foot across the wide
nursery floor, working from chair to chair for occasional
support.

"Sorra a scrap of difficulty about it," Farrell assured
him, "once you've got it firm set in your mind that it's
all an everlastin' turnin' of body into soul that's required.
All of a man's body that's nothing at all but body is nothing
but divvil. The job is to cut a good share of it right out
of you, clever and clean, an' then to inspirit the whole of
the bit you have left with all the will and force of your
soul till it's soul itself that the whole has become, or the
next thing to the whole, the way the persons that lay you
out after you die and the soul has quitted would wonder to
see the weeny scrap that was left for anybody to wake.
You could take anything that there is in the world and go
on scourin' and scourin' away at the dross it has about it
and so releasin' the workin's of good till you'd have the
thing that was nine parts body and one part soul at the
start changed to the other way round, aye and more. By
the grace of God that's the work I've been at in this place.
Half-way am I now, as you can see for yourselves, to
transformin' the body of anny slushy old drink you'd get
in a town into the soul of all kindness an' joy that our
blessed Lord put into the water the good people had at the
wedding. Nothin' at all to do but walk straight on, the
way I was going, to work the stuff up to the pitch that
you'd not feel it wettin' your throat, but only the love of

God and of man an' the true wisdom of life, and comperhension of this and of that, flowin' softly into your mind. Divvil a thing stood in me way, save only "——here the mild-hearted fanatic stooped for a moment from the heights where his spirit abode to note with a wan smile of indulgence a little infirmity of mankind's—" a few of the boys do be lying around in the bog, the way they have me worn with the fear they'd lap the stuff hot out of the tub and be killed if I'd turn me back for one instiant."

" They'll quit, from this out," the sergeant said, with immense decision. " I'll not have anny mischievous trash of the sort molestin' a man at his work."

" Ow, it's a wonderful country ! " Constable Boam breathed to himself. The words had been rising to Boam's Cockney lips at almost every turn of affairs since his landfall at Kingstown. Now they came soft and low, soft and low. A peace passing all understanding had just invaded the wondering Englishman's mind.

II

Let not the English be tempted to think that by no other race can a law be dodged for a long time without scandal. Neither the sergeant nor either man of his force was ever a shade the worse for liquor that summer. To Tom's priestly passion for purging more and yet more of the baser alloys out of the true cult there responded a lofty impulse, among the faithful, to keep undeflowered by any beastlike excess the magical garden of which he had given them the key.

For it was none of your common tavern practice to look

in at Tom's when the loud afternoon hum of the bees was declining reposefully towards the cool velvety playtime of bats and fat moths. All that plays and the opera, lift of romance and the high, vibrant pitch of great verse are to you lucky persons of culture; travel, adventure, the throwing wide open of sudden new windows for pent minds to stare out, the brave stir of mystical gifts in the heart, gleams of enchanting light cast on places unthought of, annunciatory visits of that exalting sense of approach to some fiery core of all life, watch-tower and power-house both, whence he who attains might see all manner of things run radiantly clear in their courses and passionately right. The police did not offer this account of their spiritual sensations at Tom's, any more than the rest of Gartumna did. But all this, or a vision of this, was for mankind to enjoy as it took its ease on the crumbling heaps of dry turf by the still, what time the inquisitive owls were just beginning to float in soundless circles overhead. From some dull and chilly outer rim of existence each little group of Tom's friends would draw in together towards a glowing focus at which the nagging " No," " No," " No " of life's common hardness was sure to give place to the benedictive " Yes," " Yes," " O yes " of a benignly penetrative understanding of earth, heaven, and everything else. Who such a beast as to attempt to debauch the delicate fairy conducting these mysteries? Too good to imperil, they seemed, besides, too wonderful to end. Dust, all the same, hath dimmed Helen's eyes, which seemed to so many people as if their light could not go out.

All revolutions, some pundits say, are, at bottom, affairs

of finance. And Mrs. Burke had diagnosed truly. Tom
bore within him the germ of that mortal illness of giving
away all before him. His reign in all hearts at Gartumna
resembled that of the Roi Soleil over France, both in the
measureless glory of its meridian and in the fundamental
insolvency of its afternoon. He had always given the work
of his hands, to the worthy, free and without price. The
fitness to receive was all; something sacramental about
the consumption of his latest masterpiece by small, close-
drawn parties of beautiful souls made the passing of coin
at such seasons abhorrent to Tom. "Would you have me
keep a shebeen?" he had indignantly asked, when the
sergeant made a stout, shamefaced effort to pay. So from
day to day they kept up an urbane routine, month after
month. Tom would always proffer the squat glass with a
shy, tentative gesture; this made it clear that in the sight
of God, so to speak, no such freedom had ever been taken,
or thought of, before. The sergeant would always accept
in the jocose, casual tone of a martinet making one playful
and really quite absurd exception to his rules, the case being
one which, anyhow, cannot recur, so that there need be no
uneasiness about setting up a precedent now. But all
summers end, and urbanity butters no parsnips.

The brownness of later August was deepening round
Tom's place of research before he saw that the thing
couldn't go on as it was. He suddenly saw it, about ten
o'clock one morning. That evening, when the day's tide
of civilian beneficiaries had tactfully receded from the still,
and the police, their normal successors, had laid rifle and
helmet aside, Tom held up his dreadful secret from minute

to minute while the grey moth of twilight darkened on into brown moth-coloured night He tried to begin telling, but found he couldn't trust both his voice and his face at the same time. As soon as his face could no longer be clearly seen he worked up a prodigious assumption of calm and said to the three monumental silhouettes planted black on their three plinths of turf, "I'm ruined! Apt you'll be to find me quit out of the place if you come back in two days or three."

The sergeant leapt off his plinth, levered up by the shock; "God help us!" he said. "What wild trash are you just after gabbing?"

"Me fortune's destroyed," Tom pursued. His face had crumpled up with distress as soon as he began; but the kind darkness hid that: his voice was in fairly good pre-servation. "I borrowed the full of the worth of me hold-ing to get—— " and no doubt he was going on, "get along with the work I'm at here," but felt, perhaps, that this would not be quite the thing, considering. So he broke off and said only, "The back of me hand to the Jew mortgagee that's foreclosin'."

"God help us!" again said the sergeant. "And we drinking the creature out of house and home a good while back! Men—— !" He abruptly stiffened all the muscles of Duffy and Boam with the cogent parade voice that braces standing-easy into standing-at-ease. Then he thought for a moment. O, there was plenty to think of. Tom, the decent body, put out of his farm by the sheriff. Police aid, no doubt, requisitioned. The whole district, perhaps, in a hullabaloo, like all those around it. The

14

Garden of Eden going straight back to prairie. He must be firm. " Men," he resumed, " are we standin' by to see a man ruined that's done the right thing by ourselves ? I'll engage it's a mod'rate share only of cash he'll require to get on in peace with his work. An' the three of us unmarried men, with full pay and allowances! "

The heart of the ancient and good-natured people of England aligned itself instantly with the chivalrous spirit of the Gael. " Thet's right, sawgeant," said Constable Boam.

Constable Duffy's range of expression had not the width to cover fully the whole diversity of life. He ejaculated, " Hurroosh to your souls ! Five shillin's a week."

" Sime 'ere," subjoined Boam.

" Mine's ten," said Maguire, " I've got me rank to remember."

So swiftly and smoothly may any man's business pass, with seeming success, into a small limited company. Farrell, the innocent Farrell, took heart afresh and toiled on at the disengagement of Bacchus, the actual godhead, from out his too, too solid coatings of flesh. The force stilled the first wild fears of its heart and felt it was getting good value for its money—a quiet beat for the body, and for the soul an ever open line of communication with the Infinite. Through all Gartumna a warning shudder had run at the first crisis. Now the world seemed safe again ; the civilian lamb lay down once more besides the three large lions of the law, dreaming it to be enough that these were no man-eaters. Children all, chasing a butterfly farther and farther into the wilds, under a blackening sky.

While they chased, the good old Resident Magistrate, Ponting, was dying of some sudden internal queerness he had, he that had never done harm to a soul if looking the other way could prevent it. And into Old Ponto's seat was climbing a raging dragon of what a blind world calls efficiency.

Major Coburn came, in fact, of that redoubtable breed of super-dragons, the virtuous, masterful, hundred-eyed cavalry sergeants who carve their way to commissions somewhat late in their careers. Precise as some old maids of exemplary life, as fully posted up in the tricks of the crowd that they have left as a schoolboy turned by magic into a master, they burn with a fierce clear flame of desire to make up the enjoyable arrears of discipline that they might, under luckier stars, have exercised in their youth. Being the thing that he was, how could the man Coburn fail to do harm, with all the harm that there was crying out to be done ?

He sent for Sergeant Maguire. Quin, the district inspector—quite enough of an Argus himself without extra prompting—was there when the sergeant marched into the major's room. To outward view at this moment Maguire was fashioned out of first-rate wood. Within, he was but a tingling system of apprehensions. First, with gimlet eyes the two superiors perforated his outer timbers in numerous places, gravely demoralising the nerve centres within. When these exploratory borings had gone pretty far the crimelessness of Gartumna was touched upon—in a spirit of coarse curiosity far, far from felicitation.

Maguire faintly propounded the notion that keeping the

law was just a hobby rife among the wayward natives. "They're queer bodies," he said in conclusion.

No fantasy like that could be expected to weigh with a new broom possessed with its first fine passion for sweeping. "Don't tell *me*," the major snapped. His voice vibrated abominably with menace. "You know as well as I do, sergeant, the sort of a squadron it is where a man's never crimed." He paused, to let this baleful thrust tell its tale in the agonised sergeant's vitals. Then he went on, "And you know what it means," and again he paused and the four gimlet eyes resumed their kindly task of puncturing him at assorted points.

To Maguire's previous distresses was now added the choice mortification which always attends the discovery that you have been firing off an abstract and friarly morality at heavily armour-plated men of the world. With no loss of penetrative power, the major continued, "Screening—that's what it means. Sergeants who need the stripes taking off them—that's what it means. Go back to your duty and see to it."

Sergeant Maguire withdrew.

"He'd not comperhend. He'd not comperhend," the sergeant despairfully told himself, over and over again, as he legged back the four miles to Gartumna under the early-falling September dew. If only the darkened mind of Major Coburn *could* gain understanding ! Anybody on earth, you might think, if he had any wit at all to know good from bad, must see that this was a case in a thousand —that here, if ever in man's history, the spirit which giveth life was being borne down by the letter that killeth.

But that body Coburn——! Maguire had been a soldier: he knew those middle-aged rankers. " Shut-headed cattle ! " he groaned to himself. "No doin'annything with them." The dew was quite heavy. Sundown, autumn, and all that was best in the world going the way of honey-suckle and rose. Before he reached the tin hut one of the longest in human use among melancholy's standard dyes had suffused pretty deeply the tissues of the sergeant's mind.

It seemed next morning as if that summer's glowing pomp of lustrous months were taking its leave with a grand gesture of self-revival on the eve of extinction, as famous actors will bend up every nerve in order to be most greatly their old selves on the night of farewell. Mid-summer heat was burning again, and the quicksilvery haze shimmered over the bog when Maguire went out alone to see Farrell, just as the sergeant remembered it on the day when the scorched air from the furnace first showed him the still. Farrell, a little leaner now, a little less natty in his clothes, a little more absent-eyed with the intensity of a single absorption, raised from his work the patiently welcoming face of genius called away by affairs of this world from its heavenly traffic with miracles.

" All destroyed, Tom," the sergeant said quickly. The longer he waited to bash in the unsuspecting up-turned face of Farrell's child-like happiness the more impractic-able would it have grown. " The glory," he added by way of detail, " is departed entirely."

Farrell stared. He did not yet take it in well enough to be broken.

"It's this devastatin' divvil," the sergeant went on, "that they've sent us in lieu of Old Ponto—God rest his kind soul!"

Farrell did not seem to have even heard of that sinister accession. They say there were Paris fiddlers who fiddled right through the French Revolution and did not hear about the Bastille or the Terror. Live with the gods and deal with the Absolute Good, and Amurath's succession to Amurath may not excite you.

"God help the man—can't he see he's destroyed?" Fretful and raw from a night of wakeful distress, the sergeant spoke almost crossly, although it was for Tom that he felt most sorely in all that overshadowed world.

The worker in the deep mine whence perfection is hewn peered, as it were, half-abstractedly up the shaft. Not otherwise might some world-leading thinker in Moscow have looked partly up from his desk to hear, with semi-interested ears, that a Bolshevik mob was burning the house.

The disorganized sergeant veered abruptly all the way round from pettishness to compunction. "Dear knows," he said, "that it's sorry I am for ye, Tom." He collected himself to give particulars of the catastrophe. "A hustlin' kind of a body," he ended, "et up with zeal till he'd turn the grand world that we have into a parcel of old rags and bones and scrap iron before you could hold him at all. An' what divvil's work would he have me be at, for a start, but clap somebody into the jug, good, bad, or indifferent? *Now* do ye see? There isn't a soul in the place but yourself that does the least taste of a thing that anny court in the wide world could convict for. What with you and the old priest and the new, and the old colonel below, you've made the whole of the people a very fair match for the innocent saints of God. An' this flea of a creature you couldn't even trust to be quiet an' not stravadin' out over the bog by himself like a spy, the way he'd soon have the whole set of us suppin' tribulation with a spoon of sorrow."

Farrell subsided on one of the seat-like piles of sunned peat. The fearful truth had begun to sink in. He sat for a while silent, tasting the bitter cup.

The heat that day was a wonder. Has anyone reading this ever been in the Crown Court at Assizes when three o'clock on a torrid dog-day comes in the dead vast and middle of some commonplace murder case, of poignant interest to no one except the accused ? Like breeze and bird and flower in the song, judge and usher, counsel and witnesses, all the unimperilled parties alike " confess the hour." Questions are slowly thought of by the Bar and languidly put; the lifeless answers are listlessly heard; motes of dust lazily stirring in shafts of glare thrown from side windows help to drowse you as though they were poppy seeds to inhale; all eyes, except one pair, are beginning to glaze ; the whole majestic machine of justice seems to flag and slow down as if it might soon subside into utter siesta, just where it is, like a sun-drugged Neapolitan pavior asleep on his unfinished pavement. Only the shabby party penned in the dock is proof against all the pharmacopœia of opiates. Ceaselessly shifting his feet, resettling his neckcloth, hunting from each sleepy face to the next for some gleam of hope for himself, he would show, were anyone there not too deeply lulled to observe, how far the proper quality and quantity of torment is capable of resisting the action of nature's own anodynes.

Out in the bog a rude likeness of that vigil of pain, set amidst the creeping peace of the lotus, was now being staged. Under the rising heat of that tropical day the whole murmurous pulse of the bog, its flies and old bees, all its audible infestation with life seemed to be sinking right down into torpor while Sergeant Maguire's woebegone narrative dribbled off into silence and Tom came to the last of his hopeless questions. Questions? No; mere ineffectual sniffings among the bars of the closed cage of their fate. They both lay back on the warm turf, some ten feet apart, Tom staring up blankly straight into the unpitying blue while the sergeant stuck it out numbly within the darkened dome of his helmet, held over his face, striving within the rosy gloom of that tabernacle to gather up all his strength for the terrible plunge.

The plunge had to come. The sergeant rose on one elbow. He marshalled his voice. " There's the one way of it, Tom," he got out at last. " Will you quit out of this and away to the States before I lose all me power to keep a hand off you ? "

Farrell partly rose, too. His mind had not yet journeyed so far as the sergeant's along the hard road.

" I'll make up the fare from me savings," the sergeant said humbly.

Farrell turned upon him a void, desolate face. The sergeant hurried on : " The three of us down below will

clear up when you're gone. An' we'll sling the still for
you into the bog-hole. Aye, be sections, will we. An'
everything."

Farrell seemed to be eyeing at every part of its bald
surface the dead wall of necessity. That scrutiny ended, he
quietly said, "Me heart's broke," and lay back again flat
on the peat. So did the sergeant. Nothing stirred for
awhile except the agonized quiver and quake of the burnt
air over the homely drain-pipe chimney of Tom's moribund furnace.

III

The sergeant wangled a day's leave of absence to go
down to Queenstown the day Farrell sailed for New
York. Farrell, absently waving a hand from some crowded
lower deck of the departing ship, was a figure of high
tragic value. Happy the mole astray above ground, or the
owl routed out into the sun by bad boys, compared with
the perfect specialist cast out upon a bewilderingly general
world. The sergeant came away from the quay with his
whole spirit laid waste—altruistic provinces and egoistic
alike; his very soul sown with salt. He had been near the
centre of life all the summer and felt the beat of its heart;
now he was somewhere far out on its chill, charmless
periphery—"As the earth when leaves are dead." He had
not read Shelley. Still, just the same thing.

"I've done me duty," he said in an almost God-cursing
tone as the three of them sat in the tin hut that night,
among ashes, and heard the hard perpetual knock of the
rain on the roof, "an' I've done down meself."

"Aye, and the whole of us," Constable Duffy lamented, not meaning reproach, but sympathy only; just his part in the common threnody, antiphone answering unto phone.

Constable Boam had a part in it, too: "'Eaven an' 'Ell, 'Eaven an' Ell!'"—he almost chanted his dreary conspectus of their vicissitudes. "Ow! a proper mix-up! Gord! it's a wonderful country!"

Nothing more was heard of Farrell. He may have died before he could bring back into use, beside the waters of Babylon, that one talent which 'twas death to hide. Or the talent itself may have died out in his bosom. Abrupt terminations have ere now been put to the infinite; did not Shakespeare dry up, for no visible cause, when he moved back to Stratford? All that we know is that Tom's genius can never have got into its full swing in the States. For, if it had, the States could never have gone to the desperate lengths that they afterwards did against the god of his worship.

HONOURS EASY

W HEN Colin March, a younger son of the famous
diplomatist, played in a British Embassy garden
abroad his foreign nurse gave him a tortoise.
" A useful beast," she explained : " it devours cockroaches ; they are its passion."

Colin wanted to see this beneficent passion at work. So
he captured one of the embassy's many cockroaches and
put it down in front of the tortoise's nose, like an early
Christian presented to a lion. The tortoise eyed the offered
feast, and mused deeply. The cockroach did not muse.
It was a cockroach of action. Without any apparent need
for reflection it bolted for cover, like a flash of blackness,
right into the tortoise's shell, and hid itself in that pro-
founder thinker's armpit.

The cadet of a dynasty of ambassadors was charmed with
the cockroach's wit. He filed the whole affair in a pigeon-
hole of his 'cute little mind. As he grew up he would
often chuckle to think of it. Piquant parallels would occur
to him. For a fox to go to earth under the kennels, for
landsmen to put out to sea to escape from a press-gang, for
cannon-fodder to hide at the back of the cannon—this was
the wisdom of life put into a kind of practical epigram, salt
and impudent. He adored it.

When Colin was twenty the war came. " *C'est beau,
ça,*" he said, when he saw what was done on the spot by
most of the young men that he knew. He was a con-
noisseur. He could tell a fine gesture. " It makes tools,"
he said, " of us scoffers. It is as if God had broken loose

out of the churches. Little new peers like my dear cousin Grax are becoming patrician. The rich are fairly jumping through the needle's eye—flocks of 'em—sheep at a gap."

Colin was not to be carried off his own feet by any rush to take arms. He made no holy excuses about the omission: his sense of humour saved him from that. The only kind of humbug that it would allow him to practise was humbug conscious and gleeful—not Pecksniff's humbug; only Sganarelle's. It was a vital interest to him, he demurely said, not to be dead. And how could a ruling class rule from the tomb? Might not one honestly praise Father Damien without rushing off to nurse lepers? Besides, his elder brother was badly wounded already; life, he pointed out, might at any moment become vastly more worth living than ever.

And yet the war, and the way that his caste thought about it, were not to be easily talked out of his path. Like lions, they straddled across it; like tortoises, they impended over our quick-witted cockroach. Perhaps he remembered. He seemed to. For one day he spoke to his father, his father spoke to the proper person, and Colin was given a temporary commission for "special employment" in France. From war he thus found refuge in the army. The cockroach was safe in the tortoise's armpit.

There were many strange "special employments" in France. One special employee wrote tracts upon the duty of desertion, for airmen to drop on the enemy's lines. Another kept a country house for visiting magnates to stay at. Another met dying officers' wives at the boat, and whisked them away in fleet cars to the death-beds. All these had something to do. And there were others, who may have done little harm. Colin was one of them: " G.H.Q., Fifth Echelon," was his army address. By day he sat in a tin hut, properly warmed. His casement opened on the Channel's foam. A sergeant-major brought him forms to sign, and said, " The place, sir, for the name is 'ere." He had a telephone soon—a great help in crying off dinner engagements whenever a more amusing one came. He made the right faces when anyone called. He was always game for a round of golf with the brigadier at Le Touquet. For these duties his qualifications were excellent French and Italian. He may not have borne any physical part in the great westward retreat. But how could he? What good would it do to wade into the sea?

II

For three months Colin had led a life of rude health, brightened by spirited tiffs with other saviours of the country. Ruder disputes going on elsewhere formed a dim, distinguished background for these engaging figures: battles and sieges, the Marne and the Aisne, the fall of Antwerp, the First Battle of Ypres. Colin was highly aware of the value of all this forest distance of tapestried gloom against which his own foreground figure was planted. He knew what was what. All this was romance, like the whining of winds that have blown over deserts of snow when they sniff at night round the house where a

26

person, who knows what is what, lies in bed, with the firelight leaping or musing.

Then came the Deluge. Or, rather, one of the Deluges. Colin would say : " Poor old G.H.Q. was made to be inundated and reinundated, like Holland, on proper occasions. Or it's like England, with Picts and Brythons and Angles and Normans all rushing it in their turns. All of us here are ex-conquerors, layer on layer of us. First to charge in were the War Office braves, the old hands, the mighty hunters of good jobs before the Lord. I came as a sutler with them. I'm the *jeune premier arriviste*. Then, after three months of the war, the notes of a distant recessional march strike our ears, a thunderous tramping is heard in the east, and there burst in upon us the pick of the old Regulars from the front—oh, not all of them, only those who had found that it was not the right sort of war, and that they had the right sort of friends. Approaching, they took a short run, and fairly butted and rammed their way into shelter, with all their mothers and uncles pushing them hard from behind—fell right in on top of us here with their cavalry spurs and their Guards knickerbockers and buttons and swashing and martial outsides. They were like the saints taking the Kingdom of Heaven by storm. We tried to be nice to them ; sat as close as we could to make room ; cut our work in two and gave them half, like St. Martins ; talked to them kindly and wittily They only stared. Then I knew them—the poor old Army Class worthies at school. Look at Claude, *par exemple*—his eyes ! ' There is no speculation in those eyes.' "

Claude Barbason's brain, it may be, was not all air and

27

fire. And Colin was yoked to him now. They bit wooden penholders in the same hut, and perfected a dislike of each other that they had roughed out in peace-time in London. " Ever see," Colin would pleasantly ask you, " such a good German as Claude? *Tout ce qu'il y a de Boche, absolument !* " Claude's face, indeed, with its pink-and-white gravity, heavy blue eyes and straw hair, did call up visions of some German officer prisoners. No doubt he had, through his ancestors, sojourned in England a good thirteen hundred years. Still, you could fancy him, that long ago, full of home thoughts of West Saxony, marching Londonward from the sea, mopping the sweat from pink cheeks and shaking the yellow hair away from China-blue eyes, to see for the first time, from the Kentish chalk downs, the Thames shining below through the trees.

Claude would take his own part, without positive sparkle, in this commerce in compliments. Colin to him was " the mountebank," with his " bounding cleverness " and his " beastly quotations." Colin, he said, dirtied everything that he touched; he seemed to like rolling himself and everyone else in the mud; he called the scarlet staff tabs—which they were both seeking—the Red Badge of Funk; he said the Job Lot Mess, where they and other odds-and-ends ate, ought to hang out a sign on a board, " *Au Ravitaillement des Embusqués* "; he called Fifth Echelon " Chelsea " and " Greenwich "—because, as he idiotically said, it gave a secure and honoured old age to so many young men; he was always dragging in rotten gobbets of verse, with foul under-meanings: " Soldier, rest, thy warfare o'er ' ; " Keep thou still when clans are

28

arming " ; " His tin hat now shall be a hive for bees "—
oh, there was no end to his loathsomeness !

"Speak for yourself," Claude often wanted to say.
Let the fellow befoul his own nest, and not decent people's.
What could an amateur soldier like that know about what
a real soldier must feel ? Yet Claude's scorching retorts
did not get themselves uttered—only something dry and
austere, like " I suppose we all get our orders and have to
obey them," or, " Well, if you don't mind, I'll get on with
my work. There's a war on," said with a reproving
stiffness. Then Claude would bite his penholder pretty
severely.

III

I fancy it was in the gloom following one of these
indecisive engagements with Colin that Claude's eyes were
suddenly opened, like Adam's and Eve's, and he saw that,
for the high purpose of conflict with Colin, he, Claude,
was little better than naked. If, now, he had a ribbon or
two on his bosom, all the darts of Colin's flashy, trashy wit
would be deflected ; Claude would be armour-plated, like
capital ships ; like generals, he would be able to score
without saying a word, just by sitting behind the front of
his tunic and letting it tell.

Somebody said in his hearing that night that the King
of Alania—we'll call it Alania—was soon to visit our
front. Claude listened. After dinner he cast a long,
passionate look at a framed thing that hung on the ante-
room wall. It looked from afar like a coloured plate of the
full solar spectrum, but it was labelled, " The Ribbons of

all the World' Orders of Honour." Yes, an Alanian ribbon was there—a blue one, a beauty! Fie, thought Claude, upon this quiet life in a hut, yoked with an unbeliever. Swiftly he wrote to three uncles of his—wrote as he had not written since the days when he first perceived that the trenches were no place for him.

The uncles were loyal; Claude, if a babe in some ways, was no Babe in the Wood. And they were soldiers, and well placed for doing good deeds to a nephew. One of the three was in actual charge of the plans for giving this Alanian King a good time, *vice* somebody else who was ill.

The King duly came. He was reverentially motored about from meal to meal, well in the rear of our front. And who but Claude sat in state beside the chauffeur, except when he—Claude, not the chauffeur—leapt down to open the door! In this great office Claude bore himself meekly through three dusty midsummer days. On the third evening the King and his British guides, nurses, and gillies of every degree stood somewhat self-consciously grouped at Amiens on one of that city's desolate lengths of low railway platform. The guest was going away. Abruptly the fountain of honour was turned full on, and it played in the twilight.

Nervous and kind, wishful to do the right thing by Britain, but not to keep one of France's trains waiting, the King dealt out Stars of Alania with shy expedition to all the British officers who had done anything for him. An A.D.C. stood beside him and fed the blue-ribboned trinkets into the gracious hand. Claude went in last. But even when he was bestarred three stars were visibly left

over. The King held one of them, ready to shed. The
A.D.C. was still holding an unmistakable brace. Somebody must have miscounted. Or else, as Claude came to
believe later, the Devil was in it. The fountain of honour
looked like slopping over the edge of its basin.

A little way off, in the gathering dusk, three British
officers, not attached to the King's party, were standing
—perhaps awaiting the train, perhaps not. The Alanian
A.D.C. cast a look towards them. Then he looked at the
King and drew the King's eyes towards the trio. The
King nodded. " There iss," he said sweetly, " no British
officer who iss not worthy." The red-eyed train for Paris
was now clanking out of the tunnel into the station.
" Quick, please ! " said the King, in Alanian..

The three unpremeditated vessels of the royal grace
were informed. And the angel of this annunciation was
Claude. To his unaffected distaste he found, on approach,
that one of the three was Colin. Still, Claude's not to
reason why, at any rate until later. He delivered royalty's
summons. In three minutes the three remainder Stars had
settled into their new, fortuitous homes, the King had
peace in the quickening train, and Claude had briefly let
himself go on the question of unearned increment, and was
hearing a little from Colin about the divine super-equity
of the ruling that he who had borne the burden and heat of
three full days' work in the vineyard should not receive
more than he who had wrought for one hour only.

" An hour ! " objected the literal Claude. " Why, you
only paraded for pay ! "

" Absolutely," said Colin. But he was too modest.

That dramatic scene at the station had really taken some skill and pains to bring off. Drama, they say, is the art of preparation.

Claude simmered and fumed. "Anyhow," Colin said like an angel, "it has all ended happily."

"Not that I wanted," Colin explained to me later, "this acrid Alanian blue blob. Red is the only wear, to my mind, on this obscene khaki. Still, one has to take life as it comes. So I went to the station. I even took other trouble. Why should I have to, though? Why should Claude have to eat dust all over the Pas de Calais before 'he can stick what he likes on his coat? Let's have Free Trade in all ribbons. Then they'd give real distinction. A man would write himself down just what he is by the things he'd put on. All the born base-wallahs would put up three rows on the spot, if they'd not got them now. The Samurai at the front would take care to wear nothing —they'd be like the patriciate we're getting in England at last, the fellows who won't take the peerages. I should wear dozens, but I'd be an artist about it. I'd paint like a Rubens and wear my own picture. I'd start from that deep Russian red with the bottomless lustre—the Cross of St. George, or what is it?—and fight it out in that key all the summer. Oh, I see red; I can hear it—whole chords of red, peals of it. Isn't any Grand Duke ever coming this way?"

IV

None came. But some bird of the air must soon have carried to Colin the news that a mission of British officers, heroes of Mons and the Marne, was about to visit the Russian front. For Colin wrote, swiftly and well, to the proper person in London. He had heard that for this mission ten hard-bitten fighters were needed : they had to have manners, know French, and be able to carry their wine. Colin answered the call of his country the moment he got it to come to him. It was, he saw, no case for delay. Empires perish. Before such another call came the Russians might have a republic, and no decorations about, like the poor Yanks, and then—too late, the saddest words in life, too late. Plenty of time, later on, for Colin to prosecute his conquests in France. He took his stand now with the nine other courteous and capacious linguists.

In holy Russia the primitive virtue of hospitality was so ardently practised that Colin came back crying out for a separate peace. Only a cure at Marienbad, he declared, could patch up his old body for heaven. No mere London season, he vowed, had ever made such demands upon the digestive force of the celebrants as this Muscovite joy-ride. Still, that profoundly lustrous red ribbon was his.

He brought back, besides, a lot of good stories. One was about a Japanese colonel, another guest of the Tsar's. In a Russian trench this child of the sunrise had strayed from the side of his guides and fallen in with four Russian privates. They were good lads from the country, simple but careful. They were not sure whether the war was still the old one with Japan or another. Anyhow, they considered it safer to kill a loose Jap. The faithful souls did it, and Colin declared that the consequent Russian apologies to Japan were a classic for young attachés to study, apart from their primary worth as light fiction. Richer still in comedy was the Japanese Government's plight. For, by way of good manners, it had to pretend to believe that the murder was not got up by the Tsar. And, to keep up this pose, a Russian staff captain, the guide who had not succeeded in keeping the Japanese colonel alive, had to be given a Japanese Order.

"That gaud," said Colin to the Job Lot Mess, "was a treasure, a sovereign prince of enamels. We ought to make more of the Japs. We ought to shift the whole war farther East. We might hold all the gorgeous East in fee. Churchill is right."

▼

Claude did not hear this address upon strategy. Claude too had gone East, though less far. When left alone in the hut he had thought deeply about A.D.C.'s. Peace, perfect peace, was their lot in this war. They toiled not, neither did they fight: yet honour found them; beauty fell, as it were, from the air, and was caught upon their tunics. Claude, as the New World says, figured upon it. Then he acted. Nature may not have made him expressly for action : rather, perhaps, for the contemplation of himself in some becoming light. But Colin's own devout self was not surer than Claude of the efficacy of prayer, directed to the right quarter. Had not he, too, seen the Red Sea cut in two, for his safety, and passed across dry ? And now the right quarter was clear. It was that bachelor uncle of Claude's who had lately got the command of a Corps. "Claude is descended," Colin explained, "from a long line of bachelor uncles. All Barbasons are. That's why they're so rich."

On the second day of the Battle of Loos Claude rallied round this beneficent uncle. The new A.D.C. took the place of a wild young peer who had gone mad and swindled and lied his way back to the head of an equally wild Irish platoon, then diminishing in the lost battle. The uncle told Claude about this eccentric : " Damn little fool ! I'd just been thinking of putting him up for an M.C. I hope *you'll* know when you're well off." Claude did. Like Issachar, he " saw that rest was good." And the Corps headquarters were pleasant. There would he see no enemy but winter and rough weather ; and these are not lethal in well-built châteaux. " Fritz has got a new gun," I was once told by an infantry private in that Corps, " as will carry thirty-eight miles." " Goowan," another put in, " that ain't nothing. He's got a gun would hit our Corps H.Q." The Corps, as a whole, felt sure its commander was safe.

To this essential of safety a nephew's love added several subsidiary blessings. Claude laboured, as in God's sight, to get the general's coffee milk really boiled and not just heated—a test of capacity which the general knew to be crucial. " If a man," as he said, " can't get a little thing like that right, he'll never win battles." Neither did Claude disdain to make sure that the general's boots were done just as he liked them. Of myrrh and frankincense less material Claude offered unstintingly all that his talents allowed. He was not as many A.D.C.'s are. Some of the most contumacious of men are those who do little personal things for the great. Valets to emperors, ushers to Solons, batmen to heroes—too often nothing is great to such men, and nobody either. Nearly all the most mutinous blasphemy that was talked during the war was talked in the A.D.C. rooms of the mighty. But Claude revered his chief. To him his uncle was one in whom the soul of " the real army " lived on, pestered indeed but still nobly unswamped by the rag, tag, and bobtail of Kitcheners and Territorials. Claude could feel for the Corps commander. Had not he too, for long months, endured the manners of Colin, the New Army man, in the wilderness ?

To those who knew what most people did not the uncle was known as The Derby Winner, or, more succinctly, The Crook. He had come out to France in command of a division. The first time it went into action his genius miscarried and lost a mile of ground and half his men. It was only then that the general showed his full speed. Some said he had leapt into a waiting aeroplane the moment the battle was lost. Anyway, he was in London incredibly soon, and seeing the proper person about a mark of distinction so signal that, when the bad news came in, it would look silly to turn round and give him a post on the shelf. Still, the news came, in the end ; some of these things will get out, whatever you do. Clearly The Crook was not born to command divisions. So he was given a Corps to command, and the Corps was now in the line, in a pretty hot corner, expiating this stroke of humour in high places. The Crook, however, cannot have been a wholly bad man. He must have had some natural affection. Claude had not served him a month when the uncle sent

up his name for a Military Cross. "You'll get it too,"
he told him. "Whenever these lists of recommendations
look a bit long to the people upstairs they start lopping off
names from the tail-end, and work up. So I've wedged
yours in near the middle, well up."

Claude knocked off work for the day, he was so moved.
He went to lie on his bed, to get at more leisure the feel of
the full warmth of Fortune's benediction upon him. All
things were well. Through the tall window, across the
bejewelled dewy grass of the park, he could see a white
road and troops on it: a New Army battalion—he could
tell that; they had no smartness—marching up to the
front, to go into the line, the undersized men bending
under their packs, to ease the cut of their straps on the
shoulder, and chorusing one of their contumacious songs
of mock-funk—

> Oh my ! I don't want to die;
> I want to go home.

Then the road was vacant and white for a time, till a
wailing of bagpipes arose, and a kilted battalion, dwarfed to
the size of a company, hove into sight, marching the
opposite way : four little companies like platoons, and few
officers anywhere ; the pipes skirling some fearful lament,
almost animal, like a moaning or keening of primitive
women over their dead ; the men with a stiff, savage gait
of sombre defiance—scorn of the enemy they had smashed,
of the staff that had thrown Scottish valour away, of the
non-Scottish troops that had failed on a flank, of the non-
Scottish commander-in-chief that had loosed the fool battle.

Claude was no great hand at reading that sort of print. Still, he did make out something. War was the great game; he saw that more clearly than ever; he saw, too, how great beyond all other wars was this war, how much more important a business to shine in. And, Gosh! what a facer this M.C. would be for the Mountebank!

VI

It was "at the front," of all places, that this blow fell upon Colin: not quite at the place where the smell of rotting meat hung, all that autumn, but still pretty near it —at a brigade headquarters. It counted all right for his purpose.

After his travels in Russia Colin had felt it was time he became a full G.S.O., a staff officer proper. Base was the slave who remained for ever merely "attached to" the staff. But so many Colins had felt this before, and had carried their point so completely, that scandal was feared. To avert it, a ruling had gone forth that all G.S.O.'s appointed thenceforth must have had some trench service. Poor Colin, to compass his end, had to take kit for a whole fortnight's stay in a new brigade commander's charmless dug-out, and to listen in candle-lit frowst to the banalities talked by the brigadier, brigade-major, staff-captain, and some odds-and-ends of medicine, signals, and the church. Each of these, he found, had some two things to say about life, and three jokes, so that the conversation of each was a sort of recurring decimal of five places. Each of them watched, with bitter foreknowledge, the countless revolutions of the others' antique decimals. One day, however, Colin heard a new thing.

A major, an acting battalion commander, had come in to tea. He knew the brigadier well, and, like a good soldier, he was blaspheming the great for the sweat that they will often give you for nothing. " Hardly a fortnight ago, sir," he said to his host, " just before you came to us, a Corps order came round to say some foreign devils—the Japs, I believe—had sent a wad of their ' Crosses for Valour '— sort of V.C.'s—one for the absolute ace—any rank—in each British Corps. Every C.O. was to pick out the hottest man-eater he'd got, for the Corps to select the tip-topper. I took days at the job, worried my officers, ricked my own brain with being judgmatic. I sent in a beauty at last—a sergeant. He'd got cut off in a post with five men and had held up the Boches for two days, till we got him away. He'd had a broken arm all the time—a great fellow ! All the other C.O.'s in the Corps ran their prize tigers too, for all they were worth. No good. It was all waste of time. The Cross never got past the Corps. An A.D.C. got it, a fellow just up from the base—Barton, Brabazon, Brasner—some name like that."

" The ribbon was—what colour, sir ? " Colin asked.

"Mouldy bluish, I heard," said the major, "like Stilton."

" I think the name," Colin said, " must be Barbason."

" That's right," said the major, suddenly interested in Colin. " You know him ? "

" I thought everyone did," said Colin, the man from the centre of things, almost severely. Ignorance seems, at times, as if it must almost be affectation.

39

"We're pretty provincial out in these parts," said the brigadier, softly. Colin laughed, and looked at the brigadier with new respect. Colin could take with good humour any rebuke that had wit.

"What has this beggar done?" the major vindictively asked him.

"He doesn't exactly *do* things," said Colin. "He *wears* them. Like me. Only he goes in for blue. He has just got an M.C. That's three different blues on his coat—a whole Blue Ribbon Army. No, an arrangement in blue, like a Whistler, but more chaste and natty—a glove fit of the blues. He's Little Boy Blue, and he blows his own horn." Where other people get cross, Colin becomes a few degrees more copiously vivacious.

The two seniors looked at each other. The dull plum ribbon of the Victoria Cross was the only one on the brigadier's tunic. The major had fought well in three battles, and he had not even one. The brigadier pointed to where a bar of ruddy Oriental radiance, that no Western loom could have made, glowed on Colin's breast, next to the lustrous Russian crimson. "That's pretty," the brigadier said.

"It should be," said Colin, "it's old, and Chinese. I won it by running away."

"Now, now——" The brigadier, as a colonel, had often had to curb the plunging modesty of subalterns.

"No idle boast, sir," said Colin. "Some Chinese Moltke came to G.H.Q. I was told to 'take the old euchre-player away out of this—anywhere—up to the front and get him shot over.' So we set off in a car for the front. The Far East didn't like the idea. Nor did the West. But pride ruled our will. We got down four miles from the front and walked on, up a road that felt naked and cocked up right into the air. Then the trouble came. There was a sort of *émeute* going on in the air. A flock of white puff-balls was straying about the blue sky, always advancing by having a new puff break out on ahead of it. Then something venomous fell into the road, six feet away, and hissed in a puddle—a thing like a bolt, or a nut, from the blue."

The brigadier put in a note: " A chip from one of our own Archie shells."

" No doubt, sir," said Colin. " I did not examine. I fell back. So did my lovely charge. Let nobody say a Chinese cannot run. The man who could beat me that day, to the car, must have been an Achilles. As we sped home my companion gave thanks, and made promises: first to God, then to me for my lead. He was a faithful fellow ! *Ecce signum !* " Colin touched the beautiful new ribbon on his bosom.

An orderly came in for his kit. His trench service was over. When he had gone, the two elder officers stared at each other. Simple souls abounded on our front, and near it. Else, how could we have won ?

VII

Honours, as old-fashioned whist-players say, were now easy—three all. But March had the pull in one way. For he was first to be back at the base, the honoriferous seaboard, washed by such tides as a man may take at the flood and be led on to fortune. "Always stick to the base in a war," a fatherly regular on the Q. side once advised him ; "don't be led away by love of excitement. Most of the good things go to the base at the end of a war, and most of the big chances come to it now." But Colin needed no man's help to see a church by daylight.

Colin's prestige at the seats of the mighty was rising. The Chinese hero had lauded, in august ears, Colin's daring and skill as a guide to the front. But the next call on his gifts was to be for a virtue more distinctively Christian. Appendicitis had suddenly smitten another illustrious guest of our army, a Spanish-American marshal, a neutral, and therefore more to be cherished than any ally. While he lay sick unto death in an inn at Bruay the British officer who had led him about in the time of his health, and who hated the sight of natural deaths, was telephoning all day to beg that some bedside mannerist might be sent up to carry on smoothing the pillow, *vice* himself. "The Dago only wants," he said, "a sort of Angel in the House."

Colin was offered this errand of mercy. He pondered. A long time ago he had tried for some days to learn Spanish; he might find he could talk it a little now, if he tried. And that Russian story showed how from the pure and un-polluted flesh of deceased foreign officers violet and other beautiful colours may spring. Possibly red. Red, he hoped. He looked carefully into the same coloured plate on the ante-room wall at which Claude had once tenderly gazed. Yes, both the Aureate Harvest (with Swords) and the

Bleeding Heart (with Swords, too) had red ribbons. The sick must be visited. Colin accepted.

He was back at the sea in eight days, better thought of than ever. His choice of official wreaths to put on the coffin, his turning of phrases on those little cards that are tied to the wreaths, were felt to have aided the cause. Spanish-America, too, must have felt he had done the thing well. For in due time the Aureate Harvest came in as the other kindly fruits of the earth do for the use of her Colins.

" A somewhat ghoulish business," said Claude, when this just award was gazetted. Claude, at the time, was just back from the Corps. The uncle's drafts upon Britain's man power had grown so exacting that he was transferred to a more august job, where any diminutions he made in the population of these islands would be less violently observable. Colin said he was changed from a fatal accident into an obscure mortal disease. Anyhow, he had no use for Claude any longer ; nor had Claude for him. Restored to the coast, Claude was working out a new way of fighting the Germans, a quite new engine of war. It was to have a great vogue, this new weapon. It could be used almost anywhere, except in a regiment. It was called " the reorganization of the establishment " of the department or unit of which the reorganizing person had charge.

Even in those early days the G.H.Q. heaven was one that had many mansions. They were of all sizes, and growing like melons—so fast that in some of them little time could be found for any work except settling where everybody should sit. Officers in command of departments,

43

sub-departments, and sections and sub-sections of sub-
departments would draft and re-draft plans for the further
sub-division of each into two or more parts, according to
the earliest and best biological precedents, each of the new
parts to be as important as the existing whole ; each, therefore, to be commanded by an officer of as high rank as the
existing commander of the whole ; with the natural
corollary that the existing commander should get a step
of promotion in order to have proper authority over these
branch subordinates, all of whose labours he would have
to superintend if the draft should be approved. Claude,
now a captain, was senior officer at the Sink, as his and
Colin's little department was called by its irreverent neighbours. At first the Sink had been Colin ; then Colin and
Claude ; six officers, doers of miscellaneous odd jobs, now
reposed within it ; there were a score of attendant orderlies,
clerks, and chauffeurs. Why not cut it in two and house
the two portions apart, with Claude to co-ordinate the
exertions of both, as a G.S.O. 2 and a major ? Claude fell
to work on a draft, fortified with a kind of genealogical
tree to establish the lawful descent of disciplinary power
over each of his expected twins, all the way down from
the commander-in-chief. While he was writing the draft
and drawing the tree he sometimes felt more apt to the
sword than to the pen. But he stuck to the pen gamely.

VIII

Something, I think, must have made Colin suspicious.
Perhaps he saw Claude writing with an abnormal fluency.

Or of course he may just have had a pricking in his thumbs.
Anyhow, like the intelligent dog whom God has taught
to scent in good time his master's intention to drown him,
Colin took himself off before Claude's draft was approved
by the proper person. My next letter from Colin was from
a ducal, delectable house in Mayfair :

You see, they have combed me out of the trenches.
Is it that I am seconded, or what is the term used by
you militarists ? *Tout court*, England hath need of me
here. I work in this weatherproof house, the new
Ministry of Liaison. No, I am not the Minister—only
his eyes and ears, or a portion of these, and of his understanding. A *raison d'être* for the Ministry is being prepared. Meanwhile it offers asylum to young men of
quality fleeing from Military Tribunals. Rods of the
houses of signatories to Magna Charta rush in daily and
cling to the horns of the altar. It is a dock-leaf planted
by merciful nature where the nettles grow.

It seems that some gifted Scottish statesman, out of a
job at the time, had been going up and down with a
dirk, as it were, in his stocking, till all the statesmen in
office wanted to find him something nice to play with,
lest he should stick the knife into one of their wames.
To save life in this way the Ministry of Liaison was
founded, and this man of mettle was placed at its head.
The Ministry was " to co-ordinate the functions of
various administrative departments."

Colin wrote to me later :

My reverence for this foundation grows. In this
kicked ant-heap of a Europe it must be about the last

abode of peace on earth and of goodwill towards men, all men.

> Nae German lays his scaith to us,
> We ne'er did ony harm.

To us the harried Anglo-German flies, and we make him a confidential clerk or an interpreter. Here he spends happy days of Government time in writing letters to London newspapers, mostly to " show up the Hun in our midst." Like mediaeval monasteries, we cherish through a dark and bloody age the endangered graces of life. Here, in the best types of chair, sleep the brave; here knit or crochet the fair, or, within seemly limits, carry on with the brave.

> Liaison our name;
> And I will not deny,
> In respect to the same,
> What that name might imply.

A free gallant life. To take me to the club, to lunch, my country has a car like Tennyson's full tide that " moving, seems asleep." Not a speck in the sky, except passing thoughts of what Claude may be up to. Are we not members one of another, and, if aught befall him, shall I not feel?

Under that peaceful surface considerable forces were stirring. For nearly three years our Napoleons, Alexanders, and Cæsars had been collecting merited marks of distinction. The skill of collectors had almost outgrown the supply of colligible matter. The most skilful were wearing about every ribbon there was; they had gained the whole world, and unless the world were enlarged they might as well bid an early farewell to the neighing steed and the shrill trump and go home. To keep their ardour from going the way of the former warmth of the moon, as well as for other good reasons, the O.B.E. was invented.

The new species of laurel was sent, very justly, in bulk to the Ministry of Liaison, to show how well the sender had done in creating a Ministry so worthy of reward. But Colin, through some unfortunate slip, was not on the list of those whom the King, on first thoughts, delighted to honour. He wrote to me:

They must have put me down third reserve only. The stones that these builders reject! Still, I was jammed in, head of a corner, later. Three high-stomached civilians, who work with us here, rejected the bauble, with a slight wave of the hand. I commend, I can even envy them. A civilian is free; he may guard his own honour. But to us soldiers, you know, an order's an order: mine not to make reply. In I went, third wicket down. With canine loyalty I wagged an un-offended tail, and accepted my one-third of the crumbs that had fallen from the table of the proud.

You see, Colin, to my mind, had no real humbug about him, as I understand humbug. He had not the lie in his soul. He did not tell lies to himself, nor really very many to anyone else. Humbug was Claude's special subject, not his. Colin's special subject was reds, and for that plummy red of the O.B.E. ribbon there was a place ready in his heart, or about a couple of inches above it.

The danger of being saddled with some undesirable job, as a vicegerent of Claude's, had now had time to blow over, and Colin began to hear the great wars and the tented or hutted field call him again. One month of tactful importunacy in the right place and he was gone, now in the full rosy red of a G.S.O., with the red and blue brassard of G.H.Q. too—with every guarantee, in fact, of life, liberty, and the pursuit of happiness. To begin with, he travelled from London in the uncrowded staff train, after his luncheon, instead of rising at six in the morning to catch the common, crammed leave train, like the ruck of regimental officers. These, on their way back to be killed, were carried off early to cool their heels for half a day at Folkestone till the staff train should arrive at the pier and its occupants have time to dig themselves in on all the best sites on the boat. " A kindly precaution," said Colin that night in the new mess, at Bligny, to which he found himself carried away by the car that awaited him on the quay at Boulogne : " one of the many kindly precautions we take to set the moribund free from too much love of living."

Colin found Claude commanding at Bligny. Such a find might not seem elating. But it inspired Colin at dinner that night with a fine mischievous brightness. Taking a quick look at Claude now and then, to see how the stimulant worked, Colin rattled on about that slouching file of *condamnés*, the infantry and artillery subalterns coming back from their leave, trailing up the steep gangway at Folkestone on to the deck, with all their lumpish kits on their backs, and their eternal pipes and mud-sick uniforms, and looking awkwardly round them, shy among all the seated staff people, for some solid object to sit on.

Claude rose to the fly: his face lit up a little when Colin played on his sense of the " New Army bounders' " social deficiencies. Yes, he had noticed those fellows—any one would—when he last went on leave. " Appalling crowd of navvies ! " He felt himself, for the moment, in quite warm agreement with Colin—with what he took to be Colin. Colin described to me later this tender reunion. " You should have heard him ! " said Colin. " Claude is simply so much natural, born prey for irony. He is like one of the little guinea-pigs that they give to the snakes at the Zoo. A plain shirker like me is almost decent beside him. I only deny Christ right out—I frankly skulk by the fire while He's getting crucified. Claude would sneer at the cut of the clothes that Christ wore on the Cross. We're the two thieves, but Claude is the one that sniggered. ' Bill Sikeses in Sam Brownes '—that's what the little rodent calls the fellows that he and I have deserted. ' Really rather awful,' he says—' these new officers. Quite five-sixths of them the sort of people you'd expect to touch their caps to you in civil life.' Imagine the lice on one of our Tommies finding fault with the Tommy's pedigree ! Then he remembered that he was an acting major, and also the only Regular there, so he quenched his familiar smile, like the ass in the play, with an austere regard of control."

I had never seen Colin angry before. Any common vexation only made him more gaily ironical. He may be cutting jokes in heaven yet, when Claude lies howling.

IX

So far our two heroes had had to work, if only an hour, for all they had got. The next thing to come in was, as Colin vulgarly said, a bit of a war bonus.

For months our gallant London Press had felt a painful dearth of " hero stunts " and " sob stories." Lord Jellicoe had meanly preferred the continued existence of his fleet to the proper provision of good matter for " scare heads." Heroic editors began to shake their heads over Sir Douglas Haig's want of " snap," " go," and " punch " in maintaining the daily ration of thrills for bald men in arm-chairs at home. Pending the proper measures for " gingering up " these commanders there might be some market still for a little emotionalizing about the old stunt of " the heroes of Mons." Just to fill up the gap, why not a " whirlwind campaign " for giving a special medal or star to the few living men who had stemmed the German rush upon Paris in 1914 ?

It was rather a daring piece of stop-gappery. Legends of Mons were wilting already under the first rays of the higher criticism. Irreverent people at home were comparing the casualties of 1914 with the rates at which a gifted old Regular staff had since extinguished so many New Army battalions. It had leaked out that in that infant war of 1914 there had been no bombardments that would not now seem like seasons of relative rest and release from the true, high-pressure hells of the later dispensation. Heartless rationalists had begun to reflect that perhaps it was not really so much more meritorious to join the army in times of deep peace, to get off the streets or to escape harder and grimier work, than it was to join it when join- ing meant embracing, almost at once, the fairest chance of an early death that had ever smiled on British recruits; that to the old Regular officers, in especial, the war had not brought the surrender of worldly good things, of beloved work or the plans and hopes of their lives, but a suddenly opened prospect of these happinesses—more pay, quicker promotion, a paradise of professional opportunity, and, after the one great toss-up with death in 1914, safety and ease in the odour of certified glory for most of those who had not lost the toss.

Still, it was a cheap "stunt." It needed no telegraphing: it could be "done in the office." And there was no risk of actions for libel and of "exemplary damages," as in "stunts" of detraction against great generals and admirals and ministers and leaders of industry who failed in their several ways to act up to the needs of stunt presses. So the stunt Press took courage, and ranted and gushed, boomed and bleated and shrilled. And, the War Office having no courage to take, the Mons Star was invented when nearly all the men who might have deserved it were dead. Thus do the Colins and Claudes of this world build better than they know. They fight with their backs to the walls of good bedrooms, against every foe that would take a job from them, and at the close of the long day, or earlier, some unsought meed is added unto them, besides all that they seek.

Alike in gaining this guerdon, our two pretty men were not alike in the emotions raised by its possession. Colin crowed with frank joy at the scandal. Scandals, he said, were too few in these colourless times. Scandal was only a reverberation of adventure, the fuming of timid mobs when taller spirits hustled and pushed their way through. Wherever a high plume had stirred in the world, and big throws had been made and the costly unreason of romance had been properly prized, scandal had smoked up to heaven like dust from winning chariot-wheels. This wisdom of life he imparted to Claude, adjuring him also not to misuse the new ribbon. Blue at one end and red at the other, and all shaded, watered, transitional, and connective—why, it was clearly sent by Heaven for purposes of liaison, to hitch on the garden of gentians, forget-me-nots, heliotrope, and cornflower that had first bloomed on Claude's tunic to any later sallies he might meditate in red, green, even black. Only let Claude beware of minding the foolish orders issued by poker-fed generals, men blind to the arts, as to the order in which ribbons should be arranged on the martial bosom. Imagine Velasquez or Tintoret laying his colours on in obedience to General Routine Orders! "No; blue to blue, red to red, each after its kind arrange we them."

Claude hated all such talk. Raffish gammon, only fit for a Radical hairdresser! If the King thought it good enough to honour a man, what loyal soldier would jeer or belittle? Claude had the Old Army's fine sense of relative values well lodged in his soul. Gazetted awards were no mere measures of worth: they were worth itself, crystallized, capitalized. "Perhaps it isn't easy," he said to Colin, with solemn concessiveness, "for other people to understand what these things mean to a soldier."

Somehow, Colin did not seem so confounded by this as he ought to be. Instead, he only looked at Claude as if Claude were a curious exhibit in a museum, and this wounded afresh Claude's soldierly consciousness of his own irreproachable normality. Claude felt at these times an intense and burning wish that Colin were not the son of a peer. The craving was almost physically painful, like retching. If Colin were scrubby by birth, somehow the world would seem more coherent.

But all this was an interlude. Back now to the grim realities of the war

X

War hath her triumphs of company-floating no less deserving renown than peace's. Claude had drafted and drawn to some purpose. With help from on high the Sink had become thrice itself, not merely twice, and one of these three Sinks alone was more capacious than the great original. Unhappily, Claude had not secured command of the whole trinity. Some more majestic bird of prey, a G.H.Q. colonel, had dropped like a stone out of the upper sky, somewhere close to the sun, and stuck his claws firmly into Claude's kill. But, subject always to this depredator, Claude received the fattest of the three distinct commands which owed him their being.

At Bligny he could not quite say what most of his six officers did. He felt surer about the chauffeurs. One of the officers he suspected of plans for salvaging solder from old bully-beef tins—a low job. He fancied another to be the minute early embryo of a demobilization unit. A third

c

was in unmistakable travail, writing a novel. A fourth seemed to have something to do with some of the army's visitors from abroad. A fifth moved obscurely about in a dim borderland between letters and war; rumour said that he "smuggled the dope" into papers at Amsterdam and Madrid—"a Wolff in sheep's clothing, you know," Colin told me. The sixth was said to be sure that if only the war went on long enough we should end it decisively at last by feeding all the outer world's "movies" with the right stuff, and to this happy issue it was hoped that he was making some larger contribution than anyone saw. Anyhow, they all kept moving. Cars were there for all, and petrol failed not. How was Claude to check the things they did, or left undone? He walked, unpolluted by such inquisitive cares, among the mysteries of his command.

The place itself had amenity. It was a white, classical, pre-Revolution château in a hollow between two chalk downs and beside a trout-stream. G.H.Q., with its possibly critical eyes, was safely far off. No one at G.H.Q. cared about Bligny. It was a mere trousers-button, a thing to be put out of mind until it should, hang it, come off. Claude "ran his own show"; and as it and everything in it were things without precedent there was no binding routine; he might fashion the show after his own image. And, in the course of nature, some honour accrued to every head of a show in the course of a year, and where would Colin be then?

Colin, also, asked himself that. No specific job had been given him yet. He had only been dumped on Bligny because there was no valid reason for dumping him anywhere else. There he lay down to sleep of a night and rose up of a morning to ring for a car and roll off to visit some proper person and bring to his mind the parable of the Importunate Woman. This time he had quite a long run in the rôle of that excellent female. And, like her, he got there at last. He prayed himself into a job—not, he felt, one that was quite what it should be ; still, it was not at Bligny ; it was at G.H.Q. proper, near the heart of the rose. And so Colin rolled off for good, as it seemed.

He left Claude sitting rather moodily in his " office "— the absent Comtesse de Bligny's boudoir. Claude always sat there for several hours a day. It was the bridge of his ship, and a captain looks best on the bridge. He was moody, because in these last weeks he had found that, however little a captain may do on his bridge, he may still make some sort of a mess. Two or three times he had almost had to act in some way or other, to take an absolute plunge, adopting one course and rejecting another. That was the trouble ; alternatives were like horses : he couldn't guess which was a winner. So Claude had tried hard each time to take either both courses or none, and now some captious god in the G.H.Q. heaven was not taking this so well as he might. Rumours of grand muddles at Bligny began to circulate in Olympus. " Silly little devil ! " Colin was soon to hear a dangerous brigadier-general say of Claude. " Of two evils, choose both—that's his idea."

In giving the Bligny billet to Claude the proper person concerned had not entertained extravagant hopes. Barbason, he had said at the time, wasn't a flyer ; still he was clearly fed up with the job he had had before. And the new Bligny show was too much of a pearl to cast before any New Army swine. So the proper person had hoped as hard as he could about Claude, and then had looked the other way as hard as he could, hoping no harm would come. But Claude made a truly wonderful mess of it all. What made things worse was the way that opportunities for making messes were growing. The show itself grew, as everything grew in that tropical army. More and more officers came after Colin had gone. The work of some of these had all sorts of civilian connections. Visitors came, British and foreign, some of them famous, some subterraneanly powerful, some open-eyed and quick-witted. They went away, all over the earth, telling funny stories of croppers that Claude had come in his kingdom. Colin heard some of them. Faint but disturbing echoes of horrid laughter found their way round, even to Claude. He grew angry, first, at all this vulgar demand for the unsoldierly cleverness that fools called efficiency. But he grew anxious, too, for G.H.Q. had been weak before now in the face of the howlings of beasts ; men had been thrown to the wolves. To be safe, he supposed he must get in some brainy bounder as an assistant at Bligny.

With mingled pride and embarrassment, Claude found, on reflection, how few brainy bounders he knew. Colin was much the brainiest of the few—a beast, but a clever beast. And Colin, he fancied, was only marking time just at present. Claude gallantly fought down his natural aversion, went to see Colin, and found him sparkling with health and good-humour after three hours' squash racquets. Claude set forth his proposition frankly. The trout-fishing at Bligny was not surpassing, but it was good. There was some work, it was true, but no coolie work of routine to be done on the nail, as at G.H.Q. proper, and no office hours. No generals came blowing in to inspect. And all the great people from London, who ran things, came through the place sooner or later. Colin would meet them at Bligny. A clever devil like him would put it all over them, so as to do himself no end of good.

There was certainly something in that, Colin thought. He thought a good deal. The tall ship of his warlike career had lately been lying becalmed, and heaven knew how much longer these doldrums might last. He knew that Claude had been slipping up, with good comic effect. He might slip up more, and then—yes, there were good troubled waters at Bligny; there might be a little good fishing, besides that for the trout. He temporized, and prepared.

While, for some weeks, he continued to do so, things went no better at Bligny. G.H.Q. rocked with mirth at Claude's misadventures. Pedantic precisians began to ask how this entertainment helped to beat Germany. Comic paragraphs crept into London and Paris newspapers. Somebody asked a sarcastic question in Parliament. Then at last it was felt that Claude had to go. But some of the great and the wise were sorry for Claude. They felt he had his points. Nosing civilians had hounded him out of his job. The like might happen to anyone. So the wise and the great said he should stay for a month more, and meanwhile be given a little something to make him feel better. Claude's D.S.O. was the most piquant thing in the next list of rewards for special gallantry and devotion.

A month after this piece of justice appeared in the *Gazette* Claude got his orders to hand over the Bligny command— yea, to hand it over to the forethoughtful Colin.

"You know why I'm unstuck?" Claude said to his officers when the blow fell. "Because March is a New Army man. Some ticks in the press have been blowing hot air about the 'Regular Army trade union'—saying it corners all the good jobs—that sort of bilge. So some New Army man has got to be jumped into something—any old job—just to have him ready to show. All very well if there wern't a war on. But how're we to win if they're always taking the heart out of the backbone of the army?" Claude, you see, was no great commander of metaphors, either. "I know one thing," he continued: "I'm not taking any hand in this ramp."

To keep this vow not to touch pitch, Claude, strictly speaking, did not "hand over" to Colin at all. He never explained to Colin the work of the place, the lie of the land, the things that had to be seen to. A stand had to be made against all this handing over to the unworthy, so Claude felt, and went out for the day in his car. He did this every day of the week during which he was to hand over, Colin and he being both in the house. He breakfasted early so as not to see Colin. Towards the end of one of these last breakfasts some tactless officer let fall a hint that a few tips about the routine of the show might help the new commander to vanquish the Germans.

"No! let him rip," said Claude, with an air of stern virtue. "If he slips up, all the better—show that wars are not won by sham soldiers."

58

Colin came in at the moment, and Claude left it at that, and finished his coffee inflexibly. Already his car stood at the door : another long day's service to the petrol trade had dawned. He gone, Colin frankly commended his love of the road and of his kinsfolk. " It's time he hopped round," Colin said to the rest of the table, " and talked to those uncles." No " reserve " or " discretion " for Colin. War, he said, was quite enough of a morgue anyhow, without that.

XI

For the next two months Colin was kept pretty busy. The chase engaged him by day. He rode a great many partridges down on the swelling chalk-hills, where the air and turf were divine. A sterner task was teaching an Irish retriever how to course hares. At night, any time he could spare from French billiards and bridge he employed in making up war arrears of light reading. Unto his officers he did as he would that his superiors should do unto him. " *Continuez, mes enfants*," he would benignantly say to them, whatsoever they did. After a Claude, he said, the land ought to have rest for some years. As to the men, he quoted distinguished authorities on the subject of " trusting the lads." He owned that he desired his command to have the charms of an ancient and untended garden, diverse and engaging with wayward self-sowing flowers, unmarred by the desolating militarist symmetries and uniformities of geometric " carpet " flower-beds. In short, even as Claude had trusted, Colin " ripped."

He ripped so visibly and audibly that, long before the

next harvest of decorations had time to come in, he was, even in G.H.Q.'s clement eyes, " ripe for booting," as Claude elatedly said. And booted he was, in his turn. But to boot Colin was like booting a large polygonal stone. It might hurt. Not for nothing had Colin practised all his social charms at Bligny for the last two months. Many visitors—editors, politicians, miscellaneous powers of light or of darkness—had gone back to England enslaved by Colin's little ways. One enamoured magnate had said before going away : " If any old fool in the army tries to get in your way, let me know." Colin had formed a Prætorian Guard, upside down—a little band of lusty civilians ready to hustle an army.

At Colin's cry for help his trusty bravoes fell to work like firemen. They pulled long wires, spread sinister rumours, warned proper persons, and made incipient booming sounds through certain megaphones of the Press which were known to be capable of giving forth, when in full blast, the most horrific bellowings. Nervous superior officers sought to appease Colin with long leave at home. He only used it to prime his redoubtable backers with nastier facts and more vitriolic suggestions.

You see, he was in quite a strong position : he had no regiment to be bundled back to ; his unit was " General List " ; so he might win, but he could not be smashed— he was dormy. He took up high ground—that the " Old Army Gang " had pushed him out by jade's tricks ; that, if he had to go back to civil life now with a black mark to his name, some of those wanglers must howl for it. Then he would moderate slightly this rhadamanthine tone, and would temper justice with mercy; if they had the sense to rub out the black mark, he would not be vindictive. But he must have something done to prove to the world that he had done his duty and not been turned down in disgrace.

"Oh, give him his blasted M.C. and be done with it," somebody said at last, looking up from an office table at somebody else.

"Afraid we can't, sir," said somebody else. "This new rule, you know, sir—about keeping M.C.'s for things done in action."

"Oh, damn! D.S.O. then."

"Very good, sir."

So Colin left with a new red and blue stain on his coat, and none at all on his character as a soldier. He bore no grudge against his persecutors. He was too deeply amused. What tickled him most was that he was the very last, he believed, of all the old-world D.S.O.'s, the men to whom the gaud came, like the Garter, with "no damned merit about it." A rule came into force, almost immediately after, that D.S.O.'s were not to be given any longer for telephoning or clerking, or any other mode of escape from the Germans; they were all to be for the mere hack-work of fighting.

"I'm like the last Groom of the Posset," said Colin, "or Clerk of the Royal Backstairs. I'm the Last British Wolf. I ought to be stuffed, when I die, and bought for the nation, and put up on the Horse Guards Parade. It has a kind of brain, you know, the dear old Regular Army: its madness has method. For three years of war all of it gorges itself with these D.S.O.'s—I mean, all of it that is not busy fighting. Meanwhile it invents the M.C., and settles on that also, till every unconscientious objector in France has got one or both. Then it sees all the Vandals and Goths of the New Army approaching the sanctuaries; so up goes this rule. It hath a twofold operation, like old Falstaff's drinks—keeps out any future staff crowd, and runs up the stock for the crowd who got in on the ground-floor. Look at my own little investment. Every new D.S.O. from this day forth will only prove more and more what a terrible fellow I was in the trenches."

I found that Claude, too, gave the new regulation his blessing. We met and dined, on my own way back from a leave, at the splendid new G.H.Q. Officers' Club at Montreuil. Presently the benign operation of Louis Roederer 1906 slightly unlaced and unbuttoned the fine Prussian greatcoat and boots of Claude's mind. "I seriously think," he seriously said, "that at least a large proportion of these decorations ought to be given for purely physical acts of valour. Otherwise they may lose caste, as it were, in the sight of the public. You may say to me, 'Why not just do our duty and let the public be damned?' Still, the public is there, and it's only fair to ourselves to mind, in some slight degree, what it thinks. When I got my own D.S.O. and M.C. I knew myself, and I think my friends knew, that I had earned them. But how is the public to know, or even to guess? All you can do to help it, and make the most hard-earned distinctions worth having, is to keep them connected, in people's minds, with the more obvious sorts of good work—front-line stunts and so on—the only things, I suppose, that the man in the street can understand about soldiering."

I tried to work this clear in my mind assisted by the illuminating radiance of the H.A.C. band, the foaming grape of Eastern France, and the beautiful W.A.A.C. waitresses dressed as comic-opera gitanas—all rendered curiously intoxicant by the sound of the rain on the roof and the imminence of my return to a little wet home in the earth of the salient. " So the bread of the children," I construed, thinking aloud, " ought to be sometimes— rather often—given to dogs, because somehow its being half-eaten by dogs makes it still more nice for the children ? "

" I don't call those brave fellows in the trenches dogs," said Claude somewhat distantly. He had a way of talking about the trenches, to us who lived in them, that made us feel it must only be in some incomplete, unreal sense that we lived there at all ; whereas in a spiritually higher and more valid sense he, the authentically rugged soldier, abode there himself, so that, in the sight of heaven, his were war's thorns and ours her roses.

" Nor I," said I, " really," somewhat discomfited. I scarcely ever touch irony without getting into some mess and showing up badly.

Claude, still severe, said, " And I wouldn't exactly call the staff children."

" Nor I," said I, feeling I must have been rude.

" Then I don't see your point," Claude austerely pursued.

" No ? " said I, still believing that somewhere or other

63

I had one, but not feeling quite sure. And how could one waste in ill-humour the last night of music and light, the shine of clean glasses and white tablecloths ?

XII

I next heard of Claude from Colin, in London. A portion of me had gone the way of all flesh in the salient, and Colin came to liven me up in a desolate Belgravian palace used as a hospital. Colin never grudged the War Office's time to any work of good-nature. Some men took it all for themselves.

"You've not heard about Claude ? " he answered my question. "Why, Claude has entered into his kingdom. Claude has done the impossible, the unthinkable—found a new seam, a very Bonanza, where the most piercing eyes in the army had only seen level sand. You know how all the princes and counsellors of the earth go out to visit Douglas Haig. D.H. believes they must all be longing to get sniped and bombarded, just because he likes it himself. So, in pure kindness of heart, he puts them into a car and packs them off for long days at the front. But Claude really knoweth man's heart. He has found them a way of escape—some sort of ' safety-first ' apparatus, no one quite knows what—whether it's a quiet shebeen in the wilds of the Somme where they can lie *perdus* all day till it's time to go home, or a whole dummy battlefield, well out of harm's way, with old German helmets and rifles lying about for the visitors to absorb as war souvenirs. Some brain-wave like that."

Perhaps I looked puzzled. How could a mere acting-major have so happy a thought, and no colonel or general knock him down and take it away and use it himself?

" Oh, Claude worked the flotation all right," Colin assured me. " Claude knew his chief, Blunt, was a fool—with a temper. So he unfolded his little idea at nine on a morning when old Blunt was looking his cheapest and blackest.

" ' Think it a good idea, do you ?' said Blunt. 'Well, I think it a damn bad idea, so you shall work it out yourself—and don't come whining to me when you've failed.' Can't you imagine him saying it ? "

Yes, I knew that dodge of drowning new-born reformers like kittens, in bucketfuls of detail.

" Of course," Colin said sagely, " Blunt was right in a way. If you're a downy old serpent, you don't want any infant Hercules kicking about. And yet Blunt was a fool. Claude had got him on the ground hop, just as he'd planned. For then he was able to go right on with his plan of the funk-hole. He did, and now this contraption of his is the envy of all G.H.Q. They say the proud and the great of this world are tumbling over each other to get in at the door. I hear that unless you're of royal blood, or a premier, Claude becomes quite short and dry with you. Once he had two live kings and two queens in the place, all at once—all the court cards in one hand—the sort of thing people write to the *Field* about. When he comes home from the wars he'll sell his visitors' book and buy land and live on his rents. When the august go away they always give him an order apiece before stepping into the car. His Legion of Honour is said to be lost in the crush. It's thought the ribbons will soon go all round his back, like a gym belt, in a broad band. I fear it can't last, though."

"Why ? " said I. " Are there not enough kings in the world ? "

"When a rather small hen," said Colin, " finds a large hunch of bread in the run, is a welter-weight cock to look on unmoved ? Is Blunt, because he has been a fool once, to be a fool always ? Believe me, he suffers remorse for his harshness to Claude. He will say, ' I have sinned.' He will undo the past. He will reorganize the establishment. Claude will be in the outer darkness, and Blunt will feed all the auriferous geese out of his own lily hand."

XIII

Then Colin talked about himself. He always had frankness, almost to the point of disease. He was, he said, eaten by care, because he had never yet failed to overtrump little Claude, and now little Claude had played such a whacking big trump. Colin said he had known an old woman once, in the country, who died of lying awake at night fearing the patchwork quilt of the old woman next door was getting on faster than hers. Colin avowed he was hag-ridden too, with the thought of that textile mosaic on Claude's bosom expanding swiftly and inexorably. Things, he said, must be thought out, lest he should die.

With Colin, to think was to talk ; his thought worked best along a kind of paper-chase track of vivid words laid for the pursuing intellect by the forerunning tongue. So there he sat, by my bed, and made more picturesquely clear, to himself as well as to me, the thing that had struck him most in all his war travels across Northern France between the coast and our front—how, as he went east, the ribbons on passing men's breasts seemed always to die down and wither just as the corn and the roses did, by the road, till on the wastes of thistle and poppies where the shell-fire began you would seldom see a decorated man. He thought, aloud, of that Brigade H.Q. where he had slept for a resonant fortnight—his nearest point of approach to the firing line. That thrice-wounded major there had not had a ribbon at all. None of the officers and men who had come in to that place from the actual front had had any. Colour had only begun to break forth again where, on Colin's way back to the sea, he had passed a Divisional H.Q. five miles farther west—"first streaks of auroral rose breaking out," Colin said, "only—not in the east. No stars in the east; precious little dayspring to visit it—

> " ' In front the sun climbs slow, how slowly,
>   But westward, look, the land is bright ! '

" And then it was only at Corps headquarters, twenty miles farther away from the fight, that the real noon came, all the flora of valour well out, the ' high midsummer pomp,' and so forth, fully on. The Army H.Q. again, when I got there, seemed like the tropics. I've worked it out that on the average the number of ribbons a British officer gets in this war varies in direct proportion to the square of his distance from the front. It's a ' law,' like the laws about heat and the conservation of energy."

I knew he hadn't worked it out before : he was only doing it now, led on by his own talk, that wildly intuitive advance-guard of his marching mind. And then, from ascertained facts, unquestionable laws, he went on visibly to speculate. Why should the working of any such law of nature as this—a law of the nature of man—be interrupted by any mere physical accident such as a sea ? Was it not of the very nature of things that the London clays should be to the payable sands of Boulogne and the rich quartz rock of Montreuil as these soils were to the utter deserts of Ypres and La Bassée ? All civilization, the great world movement, had always been westward. Was not human experience now confirming this scientific hypothesis ?—rain ribbons as it might in maritime France, it poured in Whitehall. That was the centre of things ; there all the fountains of honour played most freshly and amply, unexhausted as yet with watering the thirsty fields of France.

Colin left me in rather a hurry at last. He had seen a great light. Things had come clear. He had to be off and withdraw his application for fresh employment in France. Not there, but on British soil, must Claude be outshone.

The job that Colin, under this new inspiration, sought from the proper person, and presently got, was, I fancy, that of a kind of occasional A.D.C., to be lent to august foreigners passing through town on their way to visit our front. Princes and premiers, marshals and admirals, Colin saw them safe through the great wicked city. Claude might draw on them later, but Colin tapped the stream nearer its source. He had his reward. When I caught

68

sight of him next the thin red lines across his tunic had been—well, reinforced.

This narrative has to end without any climax. If it were fiction it might, no doubt, culminate in some one superlative masterpiece of acquisition by one or other of its heroes. But life does not work in that way : we constantly have to put up with the ineffectiveness of truth. The two seemed to pass out of my sight like two racing yachts on a day of light airs. First one of them would catch a little local breeze and skim away with a lead, and then would run into some patch of dead air while the other would pick up a puff and be carried ahead, to be then becalmed in his turn. I heard they looked most beautiful, with three full rows of ribbons apiece, like commanders-in-chief, and that people turned round to look at them in the street, marvelling that men so young should have had time for so much valour.

MY FRIEND THE SWAN

I

THE war had perished beyond all hope of revival; the Genius of Famine could almost be heard stalking along the corridors of our Ministries of Co-ordination, Information, and Demonstration. The long howling of wolves approaching the patent swing-doors had begun to chill the young blood of the brave and the fair upon whom war had called to warm both their hands for so long at the fire of life on those convenient premises; into the ruder lake-dwellings still to be traced by the traveller crossing St. James's Park a sense of the horrors of war—its worst, its post-war ones—was making its way. Thus may poor Pharaoh have felt in his dream when the seven lean and ill-favoured kine, "such as I never saw in all the land of Egypt for badness," came up after the seven fat kine and ate them without bulging.

Under this bludgeon stroke of fate no head that I know was more intrepidly unbowed than that which bore on its obverse side the comely and imperturbable face of Colin March. He was now twenty-five. A letter of his found me stuck at the Domhof, Cologne, the Christmas after the glorious and fatal 11th of November. Colin had written on War Office paper, the War Office being just then his strategic base and the seat of his government over cir- cumstance. He had also, somewhat regally, popped his letter into the King's Messenger's bag, distrusting the speed of the common post of our armies.

From his War Office he wrote, "I sit within this frown- ing pile, and I frown worse than it." Then I knew his spirits were high. He went on: "'Let me have war,' say I, as my friend Shakespeare says: 'It's sprightly, waking, audible, and full of vent.'" Then I felt pretty sure that he was as glad the whistle had blown as any old infantry colonel who wanted no more of his men to be chipped. He went on to mention a dear friend of ours, Claude Barbason. Being a regular, Claude would have to collapse after the war—"like other sausage balloons," Colin vulgarly wrote—from the size of a brigadier-general to that of a mere common captain. "Still," Colin added, "my mind is easy about him. As long as there is any sort of Q. side up above, as our friend the Bard says, that 'providently caters for the sparrow' Claude's rations are safe."

"Hullo!" I thought, "good deal of Shakespeare lying about!" And again, a few lines lower down, I read: "What says the downy old Swan of Avon about it?" followed by some queer quotation. Strange! Saul had come off as one of the saints. But Colin one of the pedants! No! He must have got a new game on. He must be writing like this, playing the ripe Shakespearean, to make me prick up my ears, before he let on. That would be quite in his line. He had fished in his time; he knew how to use ground bait.

II

Yes, it was a game—a beginning, as Colin said when I came home in May, of the reconstruction of Europe. It all came of a tip that had come indirectly to Colin from one who could not be wrong if the British Constitution is right. Colin's father, the old ambassador, had been there when King Edward met, at a dinner, the greatest of all the Shakespearean pundits. "Stick to Shakespeare, Mr. Bowles," the prudent sovereign had said to the perspiring student: "there's money in him." Colin had, after his fashion, figured on this. All sorts and conditions of men, he reflected, were would-be consumers of Shakespeare. All tried to quote him. Teacher and preacher and politician and trader—all of them wedged in a bit of his stuff, if they could, among their own drivel. Rightly seen, the plays were a quarry—only better: all the stone was ready cut. They were what the Colosseum had been when any jerry-builder in Rome could still go in and steal a load of Titus' or Domitian's building material. But organization was needed—the "big business" touch. No Geddes or Selfridge had come in as yet to lead the parched horse of Demand to the abounding waters of Supply and make him drink there at a reasonable tariff. A few poor shabby old tags—" The play's the thing," " Put money in thy purse," " To thine own self be true," and so on—were about all that the private consumer could put his hand on. Why, it was as if we were still only scraping a few shaley scuttles of coal, with a shovel, off the surface of Northumberland! Colin figured hard. Then he acted.

You will recall how in that summer of 1919 the fruits of what looked like a richer national culture began to load and bless our advertisement hoardings. Foker's Prime London Ales were, for the first time, recommended to us on those engaging "three-column blocks" of Autolycus singing "A quart of ale is a dish for a king." In extenuation of that mortal sin against the honour of the vine, the Golden Tagus New Australian Port, the preference of Mr. Justice Silence for " A cup of wine that's brisk and fine " was cited, with ingenious effrontery, a few weeks later. " Let me have men about me that are fat " (Froud's Fast Filling Breakfast Food) and " Not china dishes, but very good dishes " (Wild's War Saving Dinner Stoneware) were other works of Colin's first period. Like many other artists, he was to have three periods. Somebody said that these early strivings of his first period made the wall of a Tube station look as if an ounce of the stuff that excites ginger-beer had been thrown into a large puddle ; a patch, here and there, of the puddle capered and frisked feebly. That was how Colin, the ex-service man, began to feel his way back into the ranks of civil industry.

First, he had made out a list of those wise traders who advertise most. This took him the last two months of his war service. From out the unsuspecting herd, catalogued in this way, he would next mark down for the chase some veritable stag of ten, like the eponymous owner of Sprot's Spermaceti Rupture Cure. Then he would steal seductively up on the creature, holding out in his hand, as it were, a sealed packet and praising its unstated contents. Was that particular quarry aware, he would ask, that our national poet had written as if he had never had a thought in his mind except to assist in advertising the quarry's business ? Then, while the hunted thing stood spellbound, moveless as a tickled trout, Colin would swiftly explain that the Shakespeare Publicity Trust had the goods, that the fee was—for goods that were goodness itself—a mere bagatelle : and if any client would simply say he was disappointed the Trust would refund. What could be fairer ?

The bargain once struck, Colin would bring out his pearl. To Sprot, the Spermaceti Rupture Curer, he would present the prescription cited by Hotspur—

> The sovereign'st thing on earth
> Was parmaceti for an inward bruise.

For Messrs. Starr, of Dundee, the spirited authors of that cheap and unstable adhesive, North Star Gum, he would have ready the testimonial of Julius Cæsar—

> constant as the Northern Star,
> Of whose true-fixed and resting quality
> There is no fellow in the firmament.

You, a person of fastidious taste, may not think highly of these sallies. Nor, I suppose, would you think much of a Red Palmer fly for your luncheon. But that is only because you are no trout. And if you were a Sprot or a Starr you would know the makings of a good puff when you saw them. Colin knew the element he worked in. " I've only slipped up once," he said. " Remember the ' Tempest ' ? how our poor aboriginal friend said that the white settler used to pet him at first—gave him ' water with berries in't ' ? Wouldn't you say that was simply written for Jellaby's Genuine Juniper Gin ? "

No ! I wouldn't. Nor would Colin. His eyes twinkled demurely. I saw that he had been tempering business with mischief. I let him run on.

74

"No! With an absolute 'No!' the cloudy distiller, Jellaby, turns me his back. Only too much damn talk about water already, he says, ever since the war regulations drew the teeth of the stuff. I asked, did he think of a little crusade against war weakness in strong waters? You see, I might have romped in with my 'Too much of water hast thou, poor Ophelia?' Not he! Barred the whole topic. Next day—such is my rugged vicissitudinous life— I hit Partletts, the motoring overcoat people, right between wind and water—nothing but just 'There's a dish of leather coats for you,' out of the big drink in old Shallow's orchard. In art all the great things are simple. That one, and 'Tell me, where is fancy bred?' for the Viennese baker in Bond Street—Swiss now, of course—are two of my brightest particular winners."

I must have grunted.

"All very well," Colin said, "but I can't afford to stick at a pun. A poor fellow must live. As an artist pure and simple, the thing I prefer is a sweet pretty way that I have of slipping into a subject.

'When all aloud the wind doth blow,
    And coughing drowns the parson's saw,'

may not be a very subtle approach, I'll allow, to Presto's Cough Drops. But mark the airy ease of my Shakespearean glide into Stoolt's Tourist Ticket announcements:

'Talking of the Alps and Apennines,
    The Pyrenean and the River Po, . . .'

O, le grand Edouard connaissait son monde."

III

Colin had what, I fancy, must be the specific gift of the born organizer. Whatsoever his hand found to do he could find someone else to go on doing it for him, for both of their goods, while Colin went off with a light heart to conquer new worlds. He now pitched on Willan, an old Merton man who had wit and knew Shakespeare by heart but had never known how to market these wares for himself. To him Colin handed over the charge of the trade advertisement branch of the Shakespeare Publicity Trust. Willan minded the shop, at a liberal stipend; Colin's adventuring mind took once more to the road.

As he went he reflected. People who quarrelled were always hauling Shakespeare in if they could. " ' Tell truth,' as Shakespeare says, 'and shame the Devil,' " he heard some of them say, in their spite, where simpler souls would yell, "Liar!" Or else, in the line of unctuous irony, somebody's moral code was " ' more honoured in the breach,' as Shakespeare says, ' than the observance.' " Always, mark, "as Shakespeare says"; never "as Hotspur" or "as Hamlet" says. Why not show the poor creatures how to do, to some purpose, what they were now so impotently attempting? Why not found a completely new trade in munitions?

" You see," Colin said, " when a man has one of these fads that they all fight about he wants to dress it up to the nines in some suit of words or other. So he may do any one of three things. (1) He may cut out and make it a brand-new suit of his own. That's to say he may be Shaw or Whibley or Colvin or Wells, if he can—one of the top-hole polemical dragons. If so, he's of no use to me. Or (2) he may borrow, gratis, a greasy, shiny old suit of misfitting Shakespearean slop clobber, everybody else's cast-off reach-me-downs, like 'All the world's a stage,' 'The quality of mercy is not strained'—all that sort of real rag-and-bone-man's stock. Or (3) he can buy a decent Shakespearean suit, ready-made but unused, from a respectable tailor. Well, I have a mission to group number two. I'm out to lift it into group three. I'm the respectable tailor. I notice some poor anti-vivisectionist trying to clothe the nakedness of his cause in such hopelessly worn baggy trousers as those old 'quality of mercy' duds. I approach and, on suitable terms, produce from my stores the one perfect fit, the thing about Cymbeline's vivisectionist queen and her nasty games with the poison—

'Which first, perchance, she'll prove on cats and dogs
Then afterwards up higher.'

Hey? 'As Shakespeare says,' you know. You see the general line?" I did.

His pupil age, "as Shakespeare says," man-entered thus, our Coriolanus waxed like a sea in this new line of business. He would equip the sworn foes of the grape with: "'O God, that men should put an enemy in their mouths to steal away their brains!' as Shakespeare nobly exclaims." To the defence of the hard-bitten dealers in sherry he came the next week with: "'If I had a thousand sons, the first humane principle I would teach them should be to forswear thin potations and addict themselves to sack,' as

77

genial Shakespeare declares." For the vegetarian orator's use there was "Shakespeare's" significant saying, "Methinks sometimes I have no more wit than a Christian or an ordinary man has; but I am a great eater of beef, and I believe that does harm to my wits." But, lest the carnivora should be dashed, Colin had also on tap for their defenders "Shakespeare's" derisive dismissal of one member, at least, of the faction of cereals and greens: "He a captain! Hang him, rogue! He lives upon mouldy stewed prunes and dried cakes."

"No party bias, you'll own," Colin said. "Good spurs always in stock for any cock that will fight. I'm like both the lots of armourers 'accomplishing the knights,' all round, before Agincourt. That's Shakespeare too. I'm like Autolycus, selling all makes of ballad: no priggish censoring. Shakespeare again. He really *is* a great fellow; he beats the authors you buy. I've taken to reading him lately, simply for fun—of course, I couldn't afford the time to read him bang through for the business; I had to work with concordances, glossaries, all sorts of gadgets. What would you say to our starting in on theology? Great human interest, you know—men's business and bosoms. The stuff is all there, lying ready: 'Heaven is above all yet: there sits a Judge,' and so on, for the orthodox lot; the unbelieving dogs might have: 'There *is* no fellow in the firmament,' or——"

"Hullo!" I interrupted. "Heard that before."

"Ah! Apropos of the gum? What of it? Cannot a line be *dégommé*, as well as a general? Shall I not aim at concision? Shall I not crystallize? You think it profane?

78

Schrecklichheit, eh ? Ye ken na what's resisted. Out of that firmament line I might, had I not loved the Swan, have twisted a puff for the formamint merchants. Do not curb a desperate man too hard."

IV

Colin had really been reading his author. All the rest of his life he had read as little as might be, and nothing except what he liked. So the whole of his small reading was still in his head, and it amused him. Some of the men who read least can quote most, and of these he was one ; his funds of that kind might be small, but they were all in hand or at call, unlike the common bookworm's unavailable masses of locked-up capital. Several nights of diverting reading in bed had now convinced Colin that quite a lot of little games with Shakespeare remained to be played, to any agile player's advantage ; they needed only the simplest appliances, Colin assured me. "Pen, ink, a little paper, and a front of brass. No more. And in the after-time you marvel at the way that you have scored." Here I broke in with a few ribald words. There is, as Solomon hinted, a time for blank verse and a time to refrain from blank-versing.

The first of Colin's new set of games was to pull Shakespeare down in the open, take a play from him and give it to somebody else. " This game," he said, " is called the Higher Criticism. It isn't really new at all. They've played it off on the Bible till you'd suppose there was hardly a page that wasn't made up by some idle solicitor's clerk at Alexandria. Played it on poor Giorgione too, till he

has hardly a Jack picture left. Now it is Our Pleasant
Willy's turn. You'll see, in the spring publishing season.
'Love's Labour's Lost' is the loot that I go for."

"Why it?" I asked.

"Think what good manners," said Colin, "its great
people have! And their wit! They all walk about dropping
pearls, like Aladdin at court. Of course, as you know, it
only just shows that the Bard was as yet a green lad from
the Midlands. He fancied that the great *did* talk like that.
He was like our dear old Henry James when he came un-
spoilt from the States in his youth—when he hadn't found
out that the Mowbrays and Talbots of Albion call everything "beastly" or else "ripping." So down sat the bright
youth from Stratford and turned on all his own lustre and
wit; hoped he was doing his princes and peers to the life.
Then all the common people went to the play and thought
that all this richly figured stuff must be just what all the
great ones of the earth did really give off when they talked,
and that no one but they had the knack. A quaint popular
faith! And this faith is mine oyster, which I with my
pen am opening. How, I ask in the slim octavo now in the
press, could any low actor-fellow, out of the provinces too,
pick up so well the very idiom of the court? A fugitive
poacher; a holder of horses at theatre doors (you know
the silly old yarns)—how *could* the lout have had any nice
friends? Think how that must appeal to the 'better-class
suburbs,' where everyone feels in his bones that in all
essentials he is an earl changed in the cradle. Won't they
be cheered when they read me and see that Shakespeare
has got to hand over to one of themselves? To which, do

you ask ? With a graceful gesture I indicate James, eighth Earl of Pomfret. I looked that fellow up. He really seems to have been quite a brainy, presentable blood in his day— knew the chief people in France, and wrote a book about dogs, and had all the good gifts of nature."

I breathed a misgiving. Somehow I seem to be always cast for the moralist's part when I talk with Colin.

He agreed blithely. "Scrubby ? Of course. *Canaillerie ? Cabotinage ?* That's what I'm trying for. It's what we English need. We've tried the other game. We played the gentleman until we almost died of it. Lowther was much too well bred to euchre the Kaiser in Turkey. Spring-Rice wouldn't dirty his hands to do Bernstorff down in America. Same story at home. The nimble cads walk over us *paisiblement*. They have all the money, and all the papers except one or two. When the Red Terror comes they'll get all the best seats in the tumbrils. A bounding world, my friend. Bound or perish, that's what it's come to ; so good, hearty bounding for me. *Nunc saltandum est, nunc pede libero pulsanda tellus.*"

v

The book throve ; it sold well. The dryasdust scholars took pains to prove Colin a dunce. That helped it, by getting it talked of. Light-hearted readers liked seeing the dryasdusts' beards so airily pulled. A journal with the greatest circulation in the solar system (certified by chartered accountant) said that, " if not impeccable in point of scholarship," Colin was " stimulating and suggestive.'

Colin studied with an intelligent glee the anatomy of his success. He wrote to me: "'Strange job it is,' as my brother in letters, Molière, used to say, ' to amuse worthy people.' Still, I'm in so far in gammon now that I propose to go o'er. Lord, Lord, how subject we young writers are to this vice of quoting! Pardon it; the dyer's hand, you know———. But to my tale. Do you know that, ever since the two Poles were got to, there have been only two first-rate nuts left to be cracked in this world? Everest is one. The other's the Riddle, as some *crétins* say, of the Sonnets."

Oh yes, I had heard of that puzzle at school. What were Shakespeare's " sugared Sonnets " all about, beneath the sugar? Behind their foreground films of obvious meaning, the jealousies and lusts and vapours, what was the veiled import that seemed at times to be just about to break out into clearness, the way a peak does through mists, and then to recede again? What, as University Extension lecturers so love to ask, was the Sonnets' " message to *us* " ?

Colin answers the question. Just read his book. It is out, and its boom has begun. Colin narrates in burning words (Chapter I) how the revelation came to himself. Like Buddha's vision of truth, it was nocturnal. Lying in bed, on that last night before the war, and thinking of nothing at all, it seems that he suddenly saw, printed in letters of fire on the black page of night, three lines from Shakespeare's eighty-sixth Sonnet:

> Was it the proud full sail of his great verse,
> Bound for the prize of all too precious you,
> That did my ripe thoughts in my brain inhearse?

Then the vision disappeared for an instant, like one of those changing advertisement sky-signs; then it reappeared, with a change in its lettering, thus :

> Was it tHe prOud full sail of His grEat verse,
> BouNd for the priZe Of aLL too prEcious you,
> That did my Ripe thoughts in my braiN inhearse ?

Again the fiery sky-sign was turned out, and again it returned, but now all the lamps were extinct except those that had lit the twelve salient letters. Thus :

```
      H     O              H   E
  N        Z    O      LL     E
     R                          N
```

Again the switch was turned off, again a second of darkness elapsed, and again the letters of fire returned, now drawn in upon one another, like this :

HOHENZOLLERN

" Then I knew," Colin writes, in the spirit of earnest humility dear to so many buyers of books, " that the key so long sought by countless students more learned, more skilled than I had fallen into my unworthy hands. My heart gave thanks ; I leapt from my bed and ran for a text of the Sonnets. Poor George Wyndham's edition lay on my desk. He had given it me : I was his god-son. My mind was on fire. Sonnet after sonnet was rather devoured than read before the August dawn broke over Kensington Gardens below the windows of my study. And oh, the joy of reassurance ! No fiend had tricked me. The flames of those sky-signs of dream were not of the pit. They were apocalyptic. For now the revelation continued. From

page after page there shone, before my opened eyes, the
same solemn warning from that infinitely prescient mind
to our poor purblind age, so soon to go unprepared to the
slaughter because it had not understood. At one place it
was simply the darkling menace of the name William
Hohenzollern, given in full :

> ' HoW many lambs mIght the stern woLf betray,
> If LIke a lAMb He cOuld His looks translatE !
> How maNy gaZers might'st thOu Lead away,
> If thou wouLd'st usE the stReNgth of all thy state ! '

" At another place it seemed as if the saving message
had been floating over the face of the poem, as clouds drift
across a full moon, and the first two letters had not yet
crossed the rim of the bright orb. And so it remains a
clipped '-lliam Hohenzollern' for ever in the lines :

> ' CLouds and ecLIpses stAin both Moon and sun,
> And loatHsOme canker lives in sweetest bud.
> All men Have faults, and evEN I in this,
> AuthoriZing thy trespass with cOmpare,
> MyseLf corrupting, saLving thy amiss,
> Excusing thy sins moRe than thy siNs are.'

" And then, again, the poet-prophet's warning call to
his deluded countrymen would rise almost into a scream
of ''Ware Hohenzollerns ! ' in the sombre passage :

> ' WeARy with toil, I haste me to my bEd,
> For then my tHOughts, from far wHere I abide,
> IntENd a ZealOus piLgrimage to thee,
> And keep my drooping eyeLids opEn wide,
> Looking on daRkNesS.'

" Or else a yet more explicit finger would point to our particular peril in the ' 'Ware Will Hohenzollern ' of the passage, hitherto so little understood by the critics :

> ' A WomAn's face with natuRE's oWn hand painted ;
> A man In hue, aLL " hues " in His cOntrolling,
> WhicH steals mEN's eyes and women's souls amaZeth,
> And fOr a woman wert thou first created ;
> TiLL naturE, as she wRought thee, fell a-dotiNg.'

" ' Will ' ? ' Will ' ? Why, here, of course, was the simple clue to those enigmatic lines that had set the centuries scratching their heads :

> ' Whoever hath her wish, thou hast thy " Will,"
> And " Will " to boot, and " Will " in overplus.'

" What more natural way, when you come to read it with fuller knowledge, of telling our world how much more than enough it would have of the Kaiser ? That is always the way with an intricate puzzle. Once find the key and every particle of the conundrum seems suddenly to rush together into luminous unity. No obscurity now about a line that was once considered so knotty as—

> ' Think all but one, and me in that one " Will." '

" Could Shakespeare give a plainer hint to the searcher for his inner meaning ? ' Stick to it,' as we vulgar moderns would say ; ' one key opens all,' and ' Will ' is the key ; find who ' Will ' is and you'll find what I'm at.

" Last scene of all, as our great author says, that ended this epic of discovery, this tragedy of a British Cassandra not understood or regarded in time, was the disappearance of the ultimate mystery of the famous dedication, 'To the onlie begetter of these insuing sonnets, Mr. W. H.,' with its obviously ironical good wishes for such immortality as the William Hohenzollern of the 'insuing sonnets' was likely to get when once their riddle was read. What grotesque guesses the fumbling moles of literary biography and criticism have made at the man indicated by that 'W.H.'—William Herbert, William Hall, Willie Hughes, William Himself! How vainly, as Plato has said, do men hunt far and wide for the truth that was tumbling about on the ground at their feet when they started! One ray of understanding, and no mystery remained. The 'Riddle of the Sonnets' was thenceforth an insubstantial pageant faded, leaving, as our poet puts it, 'not a wrack behind.'"

VI

Now, don't go away with the notion that Colin's talent was just flimsy. A man who will take pains to ape in this elaborate way the pursy earnestness of the prosperous dealer in cheap culture cannot be a mere butterfly. Very taking smears of pathos, too, he knew how to dab on. Look at his "Foreword." There he wistfully tells how he had longed to push on with the book in 1914, when the vision was new; how the stern voice of war had called him away from the chaste nunnery of study; how, like a homing dove, he had flown fondly back to his desk the moment the other dove, the peace one, arrived at the seat of war; how well he knew now the tender emotion of Claudio:

86

MY FRIEND THE SWAN

When we went onward on this ended action,
 Ilook'd upon her with a soldier's eye,
That liked, but had a rougher task in hand
Than to drive liking to the name of love :
But now I am return'd and that war-thoughts
Have left their places vacant, in their rooms
Come thronging soft and delicate desires,
All prompting me how fair young Hero is,
Saying, I liked her ere I went to wars.

It all helped. So did Colin's brilliant idea of writing, unsigned, a furious attack on the book, in a popular Sunday paper ; the book was profoundly unwholesome, he argued ; it pandered to a morbid and sensational spiritualism only too rife in our neurotic age. This sent many to it. So, again, did Colin's great defence of himself the Sunday after—full of emotion, manly emotion. He had, take it for all in all, an excellent press. Even some critics who really knew better flirted skittishly with Colin's engaging chimera. " Almost persuadest thou me," one of them said, without a broad grin, " to be a cryptogram-marian."

I asked him how ever he came to start such a stag of a hare.

He laughed. " Hare ? *Le mot juste !* A hare did it—a very March hare, clothed all in motley—a living Baconian—one of the beauties who tell you that Bacon wrote Shakespeare's *théâtre complet* in his evenings. He came to my shop—to the controversial department, the one where we fit Davids out with smooth stones from the brook—the ballistics counter we call it. He wanted something to sling at the people who think the Swan laid its own eggs. It was the worst twister we'd had. It beat Willan himself. Before giving up I just opened the Bard at a venture, the way people used to dip into a Bible for luck when things stumped them. Blowed if the first thing I saw wasn't Falstaff yelling out, ' On, Bacons, on ! ' to buck up the robbers ! Nae shauchlin' testimony here, thought I, and I sold it over the counter to that distraught person. Then I reflected—if you can get out of Shakespeare a jemmy to help you to burgle his own house, what can't you get out of him ? That heartened me up. So I got down to work there and then, hunting for Z's in the Sonnets. Hohen—zed-ollern. You see, it was only the Z's that gave any trouble at all."

À PROPOS DES BOTTES

WAR is not what it was in the good time of Falstaff, when armies would not take the field without trains of picturesque sutlers hanging about them—sages and thieves and humorous potboys and sinister crones and debonair goddesses not inexorable to men—an auxiliary host of " character parts " who may have got in the way of the war, but did good beyond price to people writing historical novels and plays. And yet the semi-official, and even the demi-semi-official, campaigner is not quite extinct. He has turned army chaplain, or works for the Y.M.C.A., or she keeps a refreshment hut or a hospital at the base.

Of such was my friend John Macleary. He came to France and the northern bank of the Somme in 1916 as a more or less uniformed instrument of Australian kindness, bringing gift coffee, biscuits, and tea to serve to Australian troops in their very few hours of ease. He also brought, on his feet and two-thirds of his legs, a pair of top boots that stirred the imagination in man.

Leriche, our French interpreter, noticed them first. He had a nose for antiques. After saying " *Quel type !* " as the French always did after meeting Macleary, he added, " His boots, too! Something of storied, of ancient. In them I find a bloom, a fragrance of—no, I cannot tell of what age, of what dynasty."

I could not either. Fantastic in cut, fantastically unfitted for use in this of all wars, they looked even quainter than the quaint Burgundy fortifications where he and I sat and dangled our legs out idly over the castle wall of

Péronne. How was it ? Had we not known ? Was Australia not young, after all ? Or why should her boots come trailing these clouds of an uncharted glory of ancientry ?

I asked Macleary.

"Me boots ?" He flicked them sombrely with his cane. "They're nothin' to write home about. I'd as lief have your own. But, puttin' me boots to wan side for the momint, I'll tell ye a story. A poor story was it for me, an' yet it had elemints in it—"

John paused again, looking down at the mud in the moat, as a prospector might stare at a few golden gleams in a river's drift dirt. He went on :

"I had been a sheep-shearer then for eight years. Mind you, shearers make gran' money out in Australy. Ev'ry penny I saved that I could, havin' it still in me mind to go back one day an' buy th' old holdin' in Connaught, an' half the boreen, an' live on me land.

"Ev'ry year, soon as shearin' was done, an' all the boys lightin' out for the centres of civ'lization, possest with desire to shout all mankind till their cap'tal was entirely consumed, where would I be but up an' away with a start on them all, an' sneakin' guiltily into a great bank at Sydney to hoard the year's takin's, an' then out again an' away to th' ends of th' earth seekin' means of support, an' holdin' off from me the gnawing temptation to blue me whole forchune.

"Risin' a hundred an' forty pounds was it, the time I met with the Divvil. He was an Irishman too, an' a man of talent an' information. A Fellow of Trinity College, Dublin, he'd been in his time, it was said, an' the

90

greatest whisky drinker in Leinster. It was for that he was put out of Trinity, so ye may guess what he'd drink when he gave his mind to it. Other times, when he'd a mod'rate amount of drink taken, he'd talk with the tongues of men an' of angels. Write, too, he could that. He'd written wan book, an' nobody'd buy it. He must have written it either dead drunk or dead sober. If he'd have had the stren'th to put on his hat and off home, just before the third glass, an' plump down to the writin', begob he'd have had Shakespeare beat. As it was, he was just a derelic' timber-ship loose in th' Atlantic—a ruin himsilf, an' a peril to all mar'ners.

"I was wan of them, God help me. I fouled him in Sydney, an' I just steerin' me ninth golden arg'sy swiftly along a side-street to me bank, tremblin' with terror of sightin' anny friends that I ought be rights to be treatin'! I'd met him wance, or twice only, before.

"'Are ye rich, Macleary?' the pirate says, readin' me soul like a poster.

"'I am not, Brennan,' says I, God forgive me, an' I with the wealth of a Jew in me pocket.

"'D'ye *want* to be rich, then?' says he.

"'I do that,' says I, 'sincerely.'

"'It's yours,' he says, 'for the pickin' it up from the floor. That is,' he says, 'if ye have cap'tal, as I have.'

"I must have forgotten me breedin' an' stared incred'lously at the man. He that had never been known to have a coin in this world, savin' only the wan that he'd just borried off you!

" ' Macleary,' he says, ' th' innuendo is just. But I've won the two-hundred-pound prize in the Wallaby Sweep this day, an' the future lies smilin' before me. All I lack now,' he says, ' is a practical man like yourself, to keep a firm hold on his cash an' me own. The mor'l an' intellectu'l plant of the business I will supply.'

" With that he unfolded his plan. It seemed that some foreign woman in London, wan Madam Tussore, had acquired the wealth of th' Indies—that was Brennan's estimate of the profits—be keepin' a set of graven images, made up of wax—eminent burglars an' emp'rors an' all the great wans of th' earth, each in his habit same as he lived, an' admittin' the people at sixpence a time, or a shillin' itself, until they'd be awed an' entranced the way they'd be comin' next pay-day again to the booth an' bringin' the children.

" ' Think,' says Brennan, ' what poverty-stricken old sort of a pitch is London, compared to Australy ! Consider th' advantages here ! An aurif'rous soil; a simple, impreshnable white population, manny of them with incomes that rush in upon them like vast tidal waves, at intervals, same as your own, cryin' aloud to be spent; the pop'lar taste for th' arts as yet unpolluted be these pestilintial movies that's layin' waste rotten old hem'spheres like Europe; an', as if made to our hand, a creative genius like Thady O'Gorman beyant, that's the greatest warrant in Sydney for forgin' wax figures of sufferin' saints till he has all th' old women south of th' Equator weepin' tears down on to the floor of the church.'

" I'll own I was carried away be the flood of his

el'quence. Who wouldn't be ? He was gifted. An' yet I took me own part when it came to choosin' the figures that we were to start on in life. Brennan was all for Cupud an' Syky an' Bacchus an' God knows what naked old divvils besides. 'Get thee behin' me, Brennan,' says I. 'We'll come to them after. It's Kelly the Bushranger first, an' Charles Stuart Parnell and Bridget the Blessed, if I'm to go a step further. Strike well home first, to the hearts of the people.'

" 'John,' he says, 'I give in. Ye're a child of this world, an' the moment is yours. But wait,' he says, 'till the show's well afoot. Then there'll be no more holdin' the childhren of light in me person, an' that's a fair warning.'

" He was as bad as his word. But that came on after. We made a magnif'cent start in the Irish quarter of Sydney with saints an' liberators an' Manchester martyrs, an' Kelly an' Jawn L. Sullivan, all sittin' roun' a modest apartment. Brennan was almost decent about them at first. 'They'll be our Committee of Public Safety,' he says. 'To whatever a proper ambish'n may afterwards lead, we'll still keep this crowd of potboilers in perm'nent session. Null an' void are they all,' he says, 'as ingines of culture, an' yet, as strictly fiscal measures, they're good for taxin' the people.'

" Brennan did the patter wherever we opened the show. An', at that job, Envy herself couldn't say but that he was the boy that was in it. Ye know the way all the nations on earth, an' the Parthians an' Medes an' Elamites 'an the rest, are assembled in Sydney—Brennan kept them all standin' on wan foot an' rubbin' th' other sof'ly against their shins with excitement the time he'd be blatherin' on from the tribune, portrayin' Kelly's last stan' like a Grattan, an' Sullivan's fight with Gentleman Corbett, an' all th' old glories of Ireland, till I'd be worn to a thread with wantin' to hear him meself, an' I at the pay door, stemmin' the rush of the public we hadn't the room for.

"The end of the month saw us seventy pounds to the good. I, with me habits of thrift, was for bankin' the whole. But 'Put it all into the business,' says Brennan. 'That's what me poor fawther would always be sayin', that failed for the greatest sum ever known since they built the Four Courts. "Always invest business profits in your own business," he'd say, "an' keep them under your eye an' away from the claws of these railway directors an' anny wild cats of the sort that'd waste all before them." '

"That gold that we made at our first leapin' off was to be the root of all evil. Brennan's rulin' pash'n was loosed Be this and be that he sejuced me. Before he was done we'd invested every penny we'd made, an' the rest of me savin's to that, an' the leavin's out of his own unblest gains on the Sweep—invested it all in an outfit of wax monniments to celebrities nobody'd heard of—Hom'r an' Plato an' Cupud an' Syky ; Endymion, an' he bein' kissed be the Moon ; an' Antony—not the saint, bless him, but th' opposite—some loose divvil just after kissin' a hijjus black strumpet from Egypt ; an' wan Diogenysus rolled up in a tub, an' another profligut be the same name sittin' cocked up on a similar tub, but up-ended, an' he with vine-leaves dishivelled about in his hair an' an expensive glass in his hand (for Brennan must have the glass, as he said, of the period), an' God knows what other old trap-pin's of wicked improv'dence—all in wax, mind ye, an' perfectly done, I'll own that, be Thady, whose forchune we made with th' orders we gave him, an' every man-jack in the whole menag'rie the spit of the livin' orig'nal. Brennan certified that. An' he knew. He knew all, the man Brennan, barrin' the way to keep alive in this world an' not ruin all his acquaintance.

"For a few days the public endured it unblenchin.' Then the takin's fell off with a run. 'Ye've made th' old show too instructive,' says wan man, and he a good cust'mer. 'Begob, it's a ramp,' says another: 'no better vally for money than losin' your way in London an' fallin' head first into the British Musee'm.' 'Ah then, f'r shame,' says a third; 'be what I can see, the exhibuts have nothin' to do in this diss'lute Zoo that ye have, only kissin' an' drinkin'.

"Brennan was mad at that man. 'If ye'd got a clean soul in your body,' he says, 'ye'd be hove up right out of your beastly habitchul thoughts be the purity of th' artist's conciption. Be off with ye out of the booth,' he says, givin' the man what he'd paid to come in, 'an' the Divvil take ye and your thruppence.'

"If the Divvil had taken all the thruppences that we lost be these means he could have rinted a good house in Merrion Square. 'It's their fault, not ours,' Brennan would say to me mood'ly, after we'd turned out the lights on another calam'tous performance. 'No, it's not even their fault,' he'd say; 'it's the fate of th' ill-starred, stunted town-dweller to-day. They haven't it in them, poor creatures, to offer themselves, like so manny clean, empty, sinsitized sheets, to the influ'nce of art, the way anny Greek'd have done—at least, down to Per'cles. The poor nerve-worn scuts haven't got the seren'ty, nor yet receptiv'ty. They're knowin', without havin' knowledge, an' that's worst of all. We're before our time, Jawn, with this show—aye, and behind it as well. We'd have made money beside th' Ilyssus, an' we'll make it yet be the waters of Murrumbidgee, as easy as anny dog'd be waggin' his tail.'

" ' But when, Brennan ? When ? ' says I, ag'nized.

" ' When boys and girls are prop'ly taught,' says he, ' in the schools.'

" ' God help us,' says I. ' Not till then ! At the thick end, it'll be, of five hundred years ! '

" ' Mebbe,' Brennan says ; ' but wan thing ye'll have noticed. It goes deep. An artist's creation—a pome or wax figure—may seem mere foolishness to the minds of the half-idducated that think they know all before them ; an', mark you, the very same artist will find his intintions read off straight an' easy at sight be simple uns'phist'cated folk, sailors an' trappers an 'cowboys, that live with the earth an' walk be the fixed stars. It's been noticed repeatedly.'

" ' Well, an' what of it ? ' says I.

" ' Jawn,' he says, ' the right course for us is——— '

" The man's speech was om'nously quiet. That warned me. I knew I'd be bolted with in a minnut, wance the wild Piggasus of his el'quence had gathered its feet fairly under it. I interrupted.

" ' Ah, then,' says I, ' the right course for us is to

strangle Plato an' Cic'ro beyant in their beds, an' all this pernocthuous brood of vipers we have at us, eatin' our vitals, an' go back like prod'gal sons as we are to the Ranger, an' Jawn L., an' Turpin, an' anny other honest bread-winners we have in the stock, an' we sinnin' here a long time, before Hivven, an' in their sight.'

" ' Don't be too hard on the creatures,' he says, still quiet an' only holdin' me off, but I knew he was surr'p-tish'sly gath'rin' his breath for me destruction. ' Aren't they victims of circumstance, same as ourselves ? '

" ' They're a Trust,' I says, fighting hard for me holdin' in Connaught, 'a Trust of the sinistrest type of par'sites, sittin' in there on their arm-chairs an' tubs an' devourin' poor people's means of subsistence.'

" ' The men you're slightin',' he says, with the same portentous an' horr'ble calm as before, ' made this world the gran' place that it is.'

" ' Aye, an' no grander,' says I. ' An old synd'cate of wasters. Upus trees are they, a whole vin'mous grove, sproutin' above and rootin' below, an' we plantin' an' wat'rin', an' all the divvils in hell givin' th' increase.'

" ' Brennan waited an' let me anger go spendin' an' wastin' itself. Brennan was like that. He was a sthrategist —none to equal him. Then he commenced fair an' aisy, as though he were right, an' he just unravellin' all the good sense that I'd spun on him.

" ' Jawn,' he says, ' ye're practicul, in a sense. But ye should be more practicul. Semi-practicul men are apt not to apprehend the cash value of boldness. Wasn't Columbus considered, be all the Macleuries in Spain, an

97

unpracticul vish'nary—he that begot every hard-headed business man ever crawled on the face of the States? Jawn, we're the Columbuses of our age, an' a bountiful Prov'dence has dealt us, this hour, the strongest hand ever played. In goin' up country to-morra'—he dished the words out to me slowly, like stones ye'd dhrop into a well, holdin' me down with th' or'tor's eye that was in him— ' we'll be exploitin' the measureless virgin soil of the soul, tillin' intellectu'l prairies an' puttin' mor'l Niagaras to work. It's not to your love of th' ideel that I'm callin', deep as I know it to be. It's a business prop'sish'n I make when I say, let us cast our bread on the waters without, an' accept the returns. It's as men of the world that we penetrate, from to-morra, the heart of this cont'nent, committin' ourselves, an' the show, to th' auroral freshness of spirrut pervadin' a pastor'l race. They're makin' good wages, their eye is single, an' their bodies full of light.'

" I do him no justice at all, givin' you just a few scattered drops from the man's glowin' torrent of words. He was possist. He had a divvil. What chance had I in th' encounter—I that had nought but common sense on me side, an' he with a tongue that'd make a hare of Demosthins or Cic'ro or anny old hot-air merchant of all that we'd got in our three covered vans? I was beat—an' we quitted the good things of life the next day, conductin' our drove of white elephants into the desert.

" Why would I weary you with the tale? It's all in the Bible, after the Jews had quit out of Egypt, save that never an issue of manna an' quails was dished out in our wilderness—no, nor anny Christianable lan' like Palestine lay at th' end of it.

" Wance an' agen in a week of desolate travel we'd find some inhab'tants an' pitch our big tent an' call them up to be taxed. The few that'd pay at the door, in their des'p'rate thirst for an instiant's escape from the depths of the country, were soon discontented, fit to take back their money or wreck all before them. The Cath'lics objected to Henry th' Eighth an' th' Prod'stants to Brian Boru, an' all they'd agree on was hatin' Demosthins an' Plato an' Socrats.

" Wan man would say, ' What old Parliament of piddants is it ye're trying to rob people with ? I was hardly in at the door when I thought I was back at the school of the Christian Brothers at Cork, an' me with me lesson not learnt. It was a momint of ag'ny.'

" Another'd come leppin' back out of the tent an' askin' how dare we charge money for forcin' people to feel like the man in Rev'laysh'ns that lifted his head up an' saw only great beasts before him.

" ' Weed out th' old fossils,' said one friend we made, ' the nooklus ye have is quite good ; it's the stumers an' duds out of books that diloot it past bearin', the way ye'd have a millimetre of whisky destroyed with a yard of water.'

" As if we *could* weed them out at that time ! We'd have left the big tent so depopylated the people'd have said we were givin' no vally at all, good or bad. An' Brennan, in draftin' our bills, had put th' entire stress, an' all the big lett'rin', on to the blackest sheep in the pasture.

Over-cap'talized, over-tented, inflated wrong end first, we were left there between the seventeen divvils we had in our vans an' the deep ragin' sea in the minds of the people. Nothin' for it at all but to struggle on, bearin' our cross.

"We struggled till we must have come up against the Equator itself. Nothin' else'd account for the heat. We were approachin' a town at the time, an' it was of some populaysh'n. I disremember its name, but Brennan called it Sedan, soon as we set eyes upon it.

"'Question is,' he said, 'are we the French or the Germans this day?'

"Friday night it was then. We entered the place an' put up our bills for a show at two on the morra.

"Be Saturday noon the tent was ericted, the seatin' in order. I was installin' the turnstile. Brennan says, 'I'll run roun' to the vans before dinner an' move the stock into the tent, the way they'll be ready.'

"He wint. He was back in five minutes, greeny-white lookin,' an' flopped in a chair. 'Give me a drink,' he says, 'quickly.'

"'We'd been on the water-wagon, the two of us, all this long while, for the sake of the business, save an' except that I kep' a little whisky in store an' would always give Brennan wan mod'rate ration before addressin' the people. I gave him wan now, seein' the state he was in.

"'It's the heat,' he said only.

"'Whethen,' said I, 'are you queer, or what is it?'

"'Roo'ned!' he said, 'roo'ned!' an' that only.

"I had a terrible thought. Had the people broke in

before we'd provoked them, an' massacred all, the just an' th' unjust.

"'Roo'ned!' was all he kept maunderin' on.

"'Spit it out, man,' says I; 'can't ye see I'm in tormint?'

"'The creatures are meltin',' he says. 'Jawn L. Sullivan's wholely disinthegrated already.'

"'An' you sittin' there'—I let a yell at him—'an' not beltin' all roun' the town for a good fire hose, or th' ingine itself, to cool them before they're desthroyed!'

"I was leppin' out at the door, but he stopped me. 'I've tried,' he says. 'The reservoir's impty. Sorra a sup of water's left in the town. A summer-dried fountain, Jawn, when our need is the sorest.'

"He was collectin' all the force of his soul be this time. He didn't ask for more drink. He was making great efforts to find a way out of the pit we were in.

"'The ship's settlin' down, Jawn,' he says. 'It's only raft-constructin' is left to us now. Have ye anny plan ready?'

"'I have not,' says I, only feelin' the curse never fell upon Irelan' till now.

"'Then I have,' says he.

"He wint for our tin of red paint an' a brush an' then he turned wan of the old bills face down on a form an' sploshed on the back in big letters: 'Startlin' New Programme. End of the World. The Las' Day. At Enormous Expense.'

"Me rage rose agen at beholdin' that architect of disaster resumin' his life's work of layin' us waste. 'Is it lynched,' I says, ' that ye'd have us, for fraud on the people, as well as impov'rished ? ' I wished at the time I was wan of th' ancient poets of Ireland, men that'd raise blisters all over your face be their stren'th of invictive.

" He hands me the bill. ' Put it up outside the tent,' he says. An' I did. Yes, I was beat. I did what he tol' me. ' An' now we'll get busy,' said Brennan.

" Dinner an' all was forgot, in the fev'rish condish'n we were in. We carried the figures into the tent. I hadn't ever loved but a few of them. Some I had hated. But I could have cried to see them that hour Th' angel of death had breathed hard. They were all busy at it, fadin' away. Thady had played us off a cheap wax, or all wax is perfijjus.

" It was done, an' I takin' me place at the door, when Brennan stepped close. ' I'm Jacob,' he says, ' the time he went into the dark to wrestle with th' angel. A second glass, Jawn, f'r the love of God ! '

" I hesitated.

" ' It's not sensual'ty I'm schemin',' he says, very humble. ' It's this : I was born below par to th' extent of two whiskies, no more an' no less. It's only to have me mind rise to its natur'l statchure an' do its best f'r the cause.'

" I gev it him. He was speakin' the truth. ' An' now——— ' he says after.

" I open'd the flap of the door an' the people streamed in. Streamed ! A spring-tide of them. That bill had told well ; that, or Forchune had changed, we bein' down an' she a good woman.

" I looked for a riot as soon's the people should see the

102

deciption. But all I could hear was the voice of Brennan uplifted. ' Ye look, me sons,' he was sayin', ' in an amazed sort. I don't wonder at it at all. Me partner an' I have given up all to present, in this place, wan crowded hour of glorious death, the great an' the wise of all time in th' actual pangs of diss'lution, the lustrous eye dimmed and every other external organ of sense, th' ear an' nostril, dwindlin' away—peakin' an' pinin'. Nachure has helped, we'll allow. On'y in this torrid zone could anny gifts or devosh'n achieve on the stage this appallin' but chastenin' spectacle—dust returning to dust an' the pash'nate heart an' d'lighted spirit drownin' into indless night.'

" Brennan had found means to darken th' end of the tent, where th' exhibits were. In the gloom they presinted a frightful appearance. Mebbe th' obscur'ty an' horror was partly th' effect of th' or'tor's power that day, for he had a flow of words to the mouth that beat all. Th' abom'nable fix we were in had inspired the man with a ven'mous jet of heartrendin' rhet'ric. He was like Rachel, th' actress of old, keenin' the children, an' she with her own reelly lost on her.

" People's mouths kep' fallin' open. An old woman in the back seats started murmurin' softly : ' I *will* be good.' Wan or two quitted that couldn't put up with the strain anny longer. They must have told others outside the dreadful experience they'd been enjyin', for people kep' throngin' up with their money the way I was almost ashamed of the click of the turnstile in that solemn place. Brawlin' in church it felt like, an' givin' change in the temple, the way of the Jews.

" Now an' then I could catch a whiff of th' intoxicant drug of Brennan's or't'ry wafted across to the door where I stood, an' then it'd seem as if he were takin' the people up into a high place an' showin' them all the wonder an' glory of life, and then droppin' them, soft an' compash'nate, down into the depths of deprission. Broodin', wan instiant, in pity an' sorra over the perishin' hand'work of Thady. ' Empusa's crew,' I'd hear him say, ' so naked new, they couldn't stan' the fire,' an' then agen he'd fasten on to some indivijjal detail of the gall'pin' process of exterminaysh'n. Cupud's nose'd start drippin' itself on to the head of Syky, until you'd be hard set to tell which was head and which nose, the way the two figures were sold'rin' themselves into wan, an' then : ' They were lovely,' he'd say, ' in their lives, an' in death they were not divided. Think of your Shelley, me friends :

> ' All things with others' bein' mingle,
> Why not I with thine ? '

" Or else he'd skid off an' away into bursts of gran' preachin' an' proph'cy itself. Wan wad of it I remember, because of the mention of wax, was like a spaycies of hymn :

> ' Before thy breath, like blazin' flax,
> Man an' his marvels pass away ;
> An' changin' empires wane an' wax,
> Are founded, perish, an' decay.'

" And then he'd have them all tremblin' to hear his reports of the Dies Iræ itself, an' th' earth an' all shrivelled right up an' the flamin' hivvins rollin' together in wan tremenjus hol'caust of cremaysh'n. But what am I doin' at all, to attempt to give you the feel of it. It was the fearfullest thing ever auj'ence paid to go through.

"We persisted far into the trop'cal night. When the last of our patrons had quitted out of it, weighed down be thoughts of their end, Brennan flopped down on a form.

"'Unarm me, Jawn,' he says, 'the long day's work is done, an' we mus' rest.'

"I didn't attend to him at the first. 'Siventeen pound,' I said, countin' the takin's, 'all but a shillin'.'

"'Good,' says he. 'We've proved it at last.'

"'What have we proved,' says I, 'except simpletons only?'

"'We've proved,' he says, 'to-night, that intilligent men needn't put their ideels away in their pockets when caterin' for the people.'

"'We've proved they must put their hands into their pocket,' says I, 'an' take out whatever was in it. We've been like the Jews up to now, only crossin' the wilderness wance in a way. From this out, I'll engage, we'll be more like the goat in the Bible that had the fee-simple of all the Sahara.'

"But I wouldn't argue. What was the use of it? Brennan had been, all the time, a kind of deminted impostor. Firs' to las' he'd let on it was after the money he was, like a nachur'l human bein'. Under it all he'd furtively been a ravin' ideelist, aye, fit to be tied, with the pash'n he had on him for makin' people enjoy things they detisted. It must have been a kind of instinc' gone wrong. He couldn't control it. 'Twas that made him dang'rous, same as a li'ness with cubs.

"We auctioned the horses an' vans where we fell— an' little they fetched, bein' low in condish'n an' not agri-culchur'l. That and the siventeen pound we had cleared on that las' Day of Wrath was enough, an' no more, to pay off the ticks we had run on the journey up country. We re-intered Sydney pinniless."

"Not a thing left?" I asked, with mechanical sympathy.

"Not a jot or a tittle, save only "—John looked em-barrassed—" I've got on me Henry th' Eight's boots."

He flicked them again, with an air of distaste, dangling them gloomily over the castle moat of Péronne.

THE FIRST BLOOD SWEEP

And giddy Fortune's furious, fickle wheel :
That goddess blind
That stands upon the rolling, restless stone.

ANCIENT PISTOL

WHAT happened that day put me off, and the place was not quite the same to me afterwards. It was the only good dug-out I knew—I was a full corporal then—in the jerry-built front east of Bully Grenay. "If it isn't the snug little howl," McGuffin would say, "get me one." It came up to the notions I had, when a boy, of the great times a marmot must have in winter. For one thing, it hadn't been cursed with a bolt-hole—a second way out—to give you your death with the draught. That was why a Brass Hat on the prowl had once called it a death-trap. The Germans, he said, had nothing to do but to bung up the open front door with a shell-cast of earth, in order to put us to death, cheaply. Brass Hats *have* to be talking. We called the place Old Death ever after. It made one more joke against the Hats. Besides, as McGuffin said, "it's the wife ye're right fond of that ye'd be for callin' ' Y'owld toad.' "

To get in, you certainly had to be handy. You slung yourself in through a square, timbered hole, like the frame of a picture, set in the back wall of the trench. The bottom of this wooden hatch was clear above high-water mark, at any rate in good weather. Then down a dozen steps cut in the chalk. There was no need to fall down these stairs the way people did. At the foot of the stairs you did not turn right, nor yet left, as in most of the drains that passed for dug-outs in those parts. You went straight on, into the

107

heart of the land. First came a bit of clean darkness, say thirty feet long. Then a belt, thirty or forty feet thick, of the smoke that had missed the chimney-pipe over our brazier. This barrage had to be crossed. As soon as it thinned you began to get visions of lights round an altar, burning straight up and quiet. Then you were there.

That nest was right out of the world. It had the warmth and scent of the fire, and all the air there was only just tinted a dreamy blue with the smoke. We knew we had a good thing, and we kept it up well. Two waterproof sheets were always hung neatly, spread under the roof, to catch the drippings of water, and milked into tins as soon as they bellied. We had a piece of board for a larder, hanging by strings in mid-air, to diddle the rats. We always brought up from billets a little clean straw, to lay on the floor, and we were so far from the door that most of the yellow mud and white chalk had rubbed off our boots before we had crossed the smoke belt. We had saved from the outer wall of a wrecked house at Grenay a twelve-foot length of iron pipe, a down-comer, and coaxed it through the earth above the brazier. It made a grand chimney— the one that was missed, as I said, by some of the smoke. But not by too much. You could do with a frowst, in reason, whenever you came in wet from the trench.

It was a good time, too, when this trouble began. I had just whacked the rations out for the day, and the jam was strawberry. Quite half of the bully had come out Maconochie of the prime. Price, our Q. M. S. (God reward him!), had thrown in a bit of cold ham. The two men, of our lot, who were officers' servants had failed to come round, up to time, for their cigarette ration, so this was a little bonus thrown in for the rest. We were all smoking. The fire was burning all right. The post had come in half an hour ago.

It was the post, so to speak, that began it. Ince, that we used to call Coom-fra-Wigan, had started reading a paper that was all creases and curves from coming by post. I had been watching his lips working, shaping the words as he read to himself. And then he let the paper fall on his legs— of course, he was sitting down on the floor like everyone else, with his back to the wall.

" Fair puts lid on, thot do," he said in the flat, draggy way of speaking that some of them have in the north.

" Wot O ? " Lance-Corporal Mason chirped up. Mason was always great on " keeping up the *moral* o' my men," as he said.

" Ah see in paaper," Ince went trailing on, " 's 'ow foalks at whoam 'as got agaate o' stoppin' futball. Noa raacin' ! Noa bowlin' ! No whoamin' birds ! An' it's noa futball noo ! " he went on mourning.

Then Tommy Tween must cut in. Tommy would almost take the word out of your mouth. " Ow, gow it ! Turn 'em all dahn ! Never mind us. 'Ow, naow ! Wot'd we wawnt wiv a little bit of int'rest in life ? Not likely ! "

Ince went on from where he had stopped. " Ah b'lieve it's these paapers done it." He took a savage grip of the sample he had there on his thighs. " Doan't like a bit o' spoort, paapers doan't. Costs 'em mooney."

'Thet's it," Tween interrupted, as if Ince had been stealing a story of his. "The pipers done it. Want to get aht o' pyin' a fair wige to a taht for 'angin' abaht Noomawket 'Eath. It tikes a man o' skill to watch a maw'nin' gallop. Not like war correspondin'. Naow use feedin' backers a bag of emowshnal bilge abaht 'eroes an' cheery wounded an' any ol' muck. A taht must know 'is job. Ah, an' 'e's got to be there. An' wiges accordin'. You tike it from me, Wigan, it's orl a do, got up be the pipers."

Ince still gripped his paper, so to speak, by the scruff of its neck. He shook it a little. "They taalk a lot in the paaper," he droned along, "aboot 'evils o' gamblin'.' "

"Lummy!" Tween cried out. "Wot are we fightin' for? Libbety, yn't it? An' wot's backin' your ahn judgement? Libbety!"

"They sa-ay," Coom-fra-Wigan went on, "'s 'ow it taakes men's minds orf o' their work."

Tween was started now. "Puts 'em on to it, more like. Wot mide th' Austrylians tike Polygon Wood? Down't orl the worl' know they was after the ricecourse?"

"Thot's reet," Ince certified fairly.

"Sime wi' th' 'ole o' mankin'," said Tween, quite excited. It is not only that which comes in at a man's mouth that can go to his head, but that which goes out of it too. "A Chink 'll put a bit on anyfink—'orses, cocks, dogs—nuffink comes wrong to 'im. Two Chinks 'll meet in the rowd an' each tike a louse aht of 'is tunic—proper 'oppers they 'ave, sime as ours—an' 'ev a match, ricin' 'em. Ah, an' stop there all dye! An' if one of 'em finds 'e's got 'old of a flyer, 'e'll put 'im back where 'e come from. Ullow! Ullow! Not time, shorely?"

None of these orators likes interruption. And now the interrupting old world was too much with us again. First, a plash of trench water lipping over our door-step, a hundred feet off, and slopping its way down the stairs, as somebody up in the trench waded nearer and nearer, kicking the waters before him. Then the voice of the platoon-sergeant, Gort, came smashing in at the door. " Blunt and Gubbins! Get on guard."

It always seemed to me like an adventure to have my name shouted by Gort. The harsh ring in his voice strung you up, the way some kinds of cold winds do. Blunt and Gubbins were buckling their belts in a second—unbuckling your belts was all the undressing allowed you in trenches. But Gort's voice came again like a box on the ear before they were out through the smoke: " Why the hell don't you get a move on?"

Tween whispered: " Bit stuffy, the sawjint." But I could never have enough of the clang of Gort's voice. It made me feel what a lot of fine, rousing things there must be in the world that I had not thought of.

In five minutes more we heard the cascade splashing again on the stairs, and a little more than the usual swearing from somebody having the usual fight to get dry through the door. Then a few clod-hopping steps, and two men clumped in through the smoke. They were the two just relieved—he that had sworn such a lot, and another.

The other was Hanney, an eager, white boy with a

thoroughbred face, just out on a draft. He had been with us three days. He had been in the army three weeks and had seen a deal of life in the time. The first ten days, not knowing his drill, he had spoilt the good looks of a section at home. Then a sage sergeant-major had ended the bother by getting Hanney put on to the next draft for the front, where there was no smartness needed. Hanney was charmed with this bit of policy. Still, it had its drawbacks. He had not learnt the soldierly knack of letting harm miss him. His first day with us he had seen a shell fall, but not burst, on the far side of the square by the café at Bully Grenay. With the God-given thirst of a wise child for knowledge, Hanney had rushed across to examine. The shell, on second thoughts, had gone off, and there cannot have been many vacant spaces, big enough to hold Hanney, between the flying bits of metal and paving-stone and other bad solids. But he was in one of the spaces, every bit of him. This business filled him with shame. Next day he had gone on a fatigue with a score of our tin water-bottles hung on his back, to fill at the cart. By one of the flukes that do happen, a flake of a shell had sheared right through the whole of this outfit of tin-ware, turning it all into silvery ribbons and rags. It had not touched the boy. This elated the boy, for he was not to blame. Still, he was rather apologetic about the interest he took in these little occurrences. All of us, he pretty clearly thought, must have had more stirring experiences daily. So we tried to be rather nice too, and allowed the Kid had powlert up and down a bit and had two rattling days. To-day was his third. We did not know, really, whether to think that nothing could kill him ever, or that, at the rate he was going, he'd not last a week.

Tween had still a little bit left from his old rush of words to the mouth. "Wot ow!" he said to the Kid. "Is't you been gettin' aht the sawjint's shirt?"

Hanney laughed at the notion. "Shirty! He's great. I could do a lot of work for Gort and not feel tired. You didn't hear, did you, just now?" He paused for a moment. "Of course, it wasn't anything out of the way, but the old Boche put it across us a bit, just for a little. And then Gort was up and down the trench every minute, looking after us all. I'd hear him asking, before he came round the corner, 'Hanney all right there?' 'Hanney not hit?' All the time."

"Wot! Downcher knaow?" Tween almost hooted.

"I know he's all right,' Hanney said sturdily. "Regular father he is to his men." You see, Hanney was young.

It made Tween roar. "Ya silly baa-lamb! Downcher knaow Gort drew you in the sweep?"

I do say this for Tween—nobody could have thought that the Kid would mind it so much. Of course, the rest of us had long been used to the thought that whichever of us was first to be killed, each time we came into the line, would bring in a good two-pound-ten, or anything up to three pounds, to whoever had drawn the name of the deceased out of a tin hat. The queer part of it was that the Kid's joy fell right in like a soufflé.

"Was that," he said, "what we were drawing, just after I joined?"

"*You* knaow," said Tween, in a firm, encouraging

manner of speaking, like someone awaking a sleeping man for his good. " Tanner a time—'ole comp'ny in—pye aht fust pawst the paowst. Fair awticles! Wotcher draw, 'Anney, yerself ? "

" Never looked," said the Kid; " I suppose I drew something."

The way he spoke gave an evident shock to Tween's mind. The Kid seemed not to have a sense of the value of money. " Gordelpus, 'unt yer pockets, 'Anney! " said Tween, in distress. " Yer mye 'ave got on to a winner." Tommy Tween was perplexed—was the Kid a lost mind ? Or had he the wind up ? Wind, Tween seemed to conclude. "'S orl right, Kiddie," he said; "it down't do yer naow 'urt. S'pose yer do cop it fust—'s on'y like 'avin a bit to leave in yer will."

But the Kid was off on some line of thought of his own " You said Gort was sick ? " he asked us at large.

Ince admitted : " Aye—seemed a bit stooffy-liike."

" At my getting off ? " Hanney said, with a bit of a made laugh.

" Well, ya knaow, ya *do* 'ang on to life a bit," Tommy mildly reminded him.

Hanney turned with a snap on McGuffin as if Mac had taken his watch. " What were you saying to-day—about that post I was in—did you say ' hottest shop in the sector ' ? "

McGuffin tried to bluster it off. " Ach ! Hwat sort of talk are ye havin' out of ye at all, thinkin' bad of the sergeant ? "

No good. The Kid had his nose well down to the scent.

"It's Gort," he said, "is it, that fixes the post for each man?"

He was a stout-hearted kid, and he needed it, having no strength in his body at all. We could see him now, out, as it were, in the twilight, wrestling with a sizable demon. "Ah, then!" McGuffin fairly wailed in dismay, "Gort never meant it a-purpose. He's not at all the description of schamer of David, the crook in the Bible, the King of the Jews."

Ince tried to help. "A mon may be streight," he went so far as to say, "though 'e do 'ave 'is feelin's."

Then Tommy Tween launched out: "'Anney, you tike it from me—w'erever there's any bettin' people's 'earts is right. I seen it tested. Pretty 'igh too. In the trine it was, comin' back from 'Urst Pawk, an' twelve in the kerrij. Standin' up meself, an' proper tired. When we was gettin' on for Vaux'all—plice they useter tike the tickets then—a blowk gets orf 'is seat an' 'e says to me, ''Ev a seat, wowncher?'

"'Garn!' says I, 'set where y'are.' Didn' seem 'ardly worth tikin' a fyvour then, nex' door to Worterloo.

"''E was 'ot on it, though. 'Fac' is, I'm compelled,' 'e says, 'to mike an arryngement,' 'e says, 'with me creditors,' an' then 'e crep' under the seat an' stretched down to it, fur in as 'e could.

"I 'adn' 'ardly set dahn in 'is plice when a man gives a shaht from the fur end of the kerrij: 'Evens on, the c'llector nabs 'im! Evens on the c'llector!

"''Gahn 'ome, ya silly swindle,' says somebody else. 'Mike it a livin' price. We're not givin' money awye.'

" Then another blowk stawts in : ' I'll bet the dead-'ead pulls it 'orf. I'll back the dead-'ead. 'Oo'll give three to one agen the dead-'ead ? '

" Then the mawket begun to brike. ' Three to two on the c'llector,' a little quiet man stawts.

" The first man looks aht o' windy. ' Lummy ! ' 'e says, ' Vaux'all a'ready ! 'Ere,' 'e says, ' I've 'ad a good dye. I wown't be 'ard on yer. Two to one on the c'llector ! Twos on the c'llector ! I'll back the c'llector ! "

" A fair price, tike it all rahnd. We 'adn't 'ardly got our money on 'fore the c'llector come in. And now, I arst yer, Kiddie, solemn. Think 'ow easy it was for any man there, that 'ad backed the c'llector, to give that unfort'nate trav'ller awye—a lift o' the legs, a kick in th' eye to mike 'im 'oller, anyfink. Naow ! They was as fytheful to 'im as 'is backers. Kep' their 'ocks strite. Gev all the cover they knew. Naow winkin' ! Naow grinnin' ! Naow nawrsty tricks to rise suspicion ! Own'y think of it, 'Anney. 'Yn't you reelly sifer in th' 'ands of a 'ot sport as stands to win a bit if you're done in than wot you'd be in th' 'ands of 'most anyone else ? 'Struth. You think it owver, Kiddie, 'fore you gaow on guard agen. You got three hour."

" One," said the Kid. He looked at a pretty wrist watch that he had. " No. Thirty-five minutes."

" 'Ullow ! " said Tween, sharply. " 'Ow's thet ? "

" I swopped a turn with Banks."

" Sawjint senction it ? " Tween asked, sharply again. You see, Tween was really a fool. He had just worked like a good one, telling the tale, to keep up Hanney's heart, and now he would undo all the good, as likely as not, by a word let slip in a hurry, putting the Kid up to notions again. We could see it act on Hanney directly; he turned sharp on Tween, as if to pick something out of his face. That made us feel awkward; so there was a bit of a silence

When you don't talk you can listen. One of the good points of Old Death was that the sound of shells would come to you there a deal bigger than life, and yet muted, as if every burst were a marvel, and yet far away, and nothing to us, like a wonderful thing in a book. We could hear it the way a child hears the big waves when he is in bed in a room on the land side of a house by the sea. At each burst the earth round us thudded softly; it did not seem to feel much of a shock—only a muffled, dreamy sort of heaving, as if it were not sleeping well. But there was always a kind of pulse, slow or fast, in this rumbly noise, and now it was rising.

" Un'ealthy weather, up top," said Tween, as if that could do any good.

And then Pratt, that had not been minding our talk, but deep in one of his endless snarly games of cards with Barnard, away in a corner, broke in like an idiot with " Sawjint Gort aht fancy snipin' agen, drawin' enemy fire ? "

Hanney stared at him too. " More evidence—eh ? "— he didn't say that, but the thought was plain in the eyes of the Kid. McGuffin saw it there, too, for he charged in to give a lift with the job where Tommy had left it. You see what we were at. We were out to make the Kid fancy that human nature was lovely.

"That's a pow'rful case that ye've cited, Tommy," McGuffin began, "but here now is one that beats all. I know, for I see ut meself, in Gallip'li. Hot summer it was at the time, an' manny dead Turks a long time in the open, the way they'd be swelled beyant your belief, like the dead transport horses below be the road. So what'd we do, to be passin' the time, but each adopt a Turk for his own an' call him Hassan or Achmet or divvil knows hwat an' back him to burst before annyone else's. Apt we'd ha' been, but for that wan divarsion, to die wi' th' ongwee of bein' definselessly et be the flies. Now, there wasn' a night but wan or more of ourselves 'd be goin' out wirin' or bombin' or only collectin' a good souvenir. An' never a case—mind ye, never as much as wan case—of a man attemptin' to mek himself rich be deflatin' the Turk he'd fancied! Not an offer at it! If it had been a matter of wanglin' a share of efficiency pay out of th' army we'd all have been out afther dark, puncthurin' every balloon in the lan'scape. 'Twas th' instinc' of sport kep' us straght. Men that had nothin' at all between them an' consid'rable wealth but givin' the touch of a bay'net in passin' ud dash the cup of joy down from their lips, the aquals of suff'rin' saints."

The feast of beauty was spread, and we all looking to Hanney to fall to and eat and be comforted. But he hardly touched it at all; and that little, I think, for civility only. The Kid was always polite about anyone's yarn. I thought of how Isaac may have behaved when he saw what his father was up to, and how the boy looked while the old man was laying the fire. The pulse of the rumble around us was going on rising.

All we could do was just to get on with the little collection we had been making for Hanney. But I had no yarn about me, to put in the hat. Ince had none either, and yet he tried to put a mite in. "Thot's reet," he piped up like a man, "Saame all worl' oaver. 'Indoos, coons, Chinks, all t'saame—nobbut fair do's anywhere, come to spoort. Tha's 'eard 'ow a Chink's buried, 'Anney ? Noa ? Dropped fair into t'graave, any'ow, saame as toss wi' cricket-bat, an' if 'e pitches faace oop it's kingdom coom. An', faace down, 'e's booked t'other gaate. Soa Chinks saay."

"Thet's right. Thet's religion, thet is," Tween corroborated.

"Eh, but think wot it ud be"—Ince ground it out, very slow and serious—"if it were all lef' to religion. Ma woord, ye'd 'ave a proper ramp soa's ye couldn't 'ave noa confidence—priests and oondertaakers queerin' toss every tiime, to maake a bit from t'relaations, 'cordin' t'wheer they'd liike t'felly to goa. Saaved by spoort, thot's wot 'tis. Every Chink in paarish 'as a bit on too, so's if paarson did t'dirty on 'em 'e'd be tore to bits by t'losers. Keeps t'clairgy streight, do thot."

"One for the dad, that!" the Kid said to me, low, when he had heard Ince out to the end, very politely. I had not known the Kid's dad was a parson. Then he got busy again on those thoughts of his own. "What did you say?" he said, "is the name of this hole?" and I had to tell him " Old Death," or " The Death Trap." That did him no good, I could see. And then Barnard, the rough of the place, must come butting into our talk.

Why couldn't Barnard have just carried on with his growling at Pratt and spitting about all over the floor ? It was temper, I think. He always spat brown the first two days in the trenches, and when he spat white, about the third day, we knew what his temper would be until we came out of the line and he got at the beer. " Paasons," he came creaking in, " is t'worst soart. Ah seen a bit o' paasons—ah, an' Chinks too—'s well as you. 'Ad charge of a gang o' Chinks, ah 'ave. Maakin' a dam. An' Scowfiel', the ganger nex' sector to me, 'ad been a church paason. They'd give 'im the go for likin' 'is beer. Soa, when a Chink died, in 'is gang or mine, we'd 'ave a bit on, saame's you was tellin'. One daay 'e come roun' to me. ' Bung Wun,' 'e says, ' is gone west. 'Ave ye any sort of a fancy about 'im ? '

" Says I, ' 'E was a laad. The first time I seen 'im, Bung 'ad took a good 'old o' both a friend's 'ands in 'is teeth an' was poonchin' 'is friend's two ears, fair an' easy. A gradely feighter ! 'E'll be for glory.'

" ' I doubt it,' says Scowfiel'. ' Bung 'ad the craf' an' subtlety of the Devil, swingin' the lead himself an' teachin' that sin to the young till the doctor'd not know were they shammin' illness at all or had they some wil' disease out of Asia he hadn't got down in his books. Nowt lef' for the doc but to ration 'em two days o' sickness a month, every man, an' pay stopped at t'first word of a tummy-ache over the ration. An' then Bung did 'im down, by formin' a comp'ny for underwritin' losses of pay. Bung's for 'Ell, take my word, if there's any moral rule over t'universe.'

" ' Ye'd bet that ? ' I asked. I liked my own opinion as well as another's.

" ' I would,' says he. ' A quid, level.'

" I took 'im.

" Firs' thing's I saw at t'funer'l was 'ow t'bearers couldn't 'ardly 'ol' up under t'corpse. Nex' was 'ow t'corpse fleed through air into t'graave soon as t'bearers let goa. Fair shook, t'earth did, when Bung landed. Then ah looked into t'graave an' see Bung lie on's faace an' bes' part of a 'oondredweight o' scrap iron busted oot through t'front of his trousies. Scowfiel' thot was. Scowfiel' done thot. E'd been a paason."

Pratt had dealt out the cards. " Eh, but they're poppin' 'em in, oop top," Barnard said, as he took up his hand. Little he cared. He was the man due to turn out for guard in twelve minutes with Hanney. But he was the kind of ox that will crop the weed grass in the butcher's yard. The rest of us listened, or felt the Kid listen. All through the earth round us the long, blunt rollers of sound were undulating and heaving more and more swiftly. They had been rather like distant thunder heard across plains; now they were more like thunder heard among mountains, where each peal lasts on into the next.

Hanney took the look at his wrist that he would not take while Barnard was pitching his tale. It made us want to darn the hole Barnard had ripped in the web we had been trying to weave, as you might say, across the

Kid's eyes. Before we could think how to do it five minutes had gone, two hands of cards had been played; Barnard was prodding on Pratt to another deal. "Dish 'em oot, mon! Gawd's saake, dish 'em oot! Owceans o' tiime!"

Pratt dealt, Barnard keeping a savage look-out on Pratt's hands for a foul, and marking time with a grouse or two at the bursts that were quickening outside. "'Ark at 'em!" he snarled, and, again, "An' we settin' 'ere in our misery!" Barnard said it the way you may hear a buck beggar practise a whine. He had no love or fear of man or God, death or dishonour; he was a kind of brave cur, base and fearless, that all soldiers know.

Ince was still fumbling about for some way to give a good turn to the talk. "An' what did t'Chinks saay," he asked, "to your Scowfiel'?"

"Chinks! Barnard jeered. "The silly swine. They was 'eartbroaken. 'Twasn't t'mooney. Moast all of 'em was winners. But they was 'eartbroken. They'd thowt as Englishmen were sports an' not like Japs as oughter all be warned off toorf, ex-offishyer, saame daay as they're born. Fair broaken-'earted, t'Chinks was. The silly sof's!"

"Playin' the game is playin' the game, all the world over," Corporal Mason put in. He would have made a great curate.

Nobody minded him, any more than if he were really a preacher. Hanney's face, I thought, was getting still greener. Pratt had dealt, and was looking very old at his cards. They must have been bad. "Serve the Chinks bloody well right, an' you too," he said to Barnard nastily.

THE FIRST BLOOD SWEEP

" Didn't this Scahfiel' o' yourn tike 'is chawnce o' you
plyin' orf the ballast on 'im, in the seat o' Wun's slacks
'stead o' their front ? 'Yn't you to tike a bit o' risk as well
as 'im ? 'Ynt 'e to show a bit o' talent ? Ah, would yer ? "

Pratt had spat out the last three words like the swear
of a cat. He must have thought Barnard was trying
it on. Perhaps he was. The two of them held their
tongues for a minute, and minded their job, while Tween
tried to pick up an odd bit of good out of the mess they had
made. " O' course," he began, " yer mye get a Scahfiel'
or two in any ol' country."

Pratt had lost, by this time, both money and temper.
" Country ! " he squealed. " Gordelpyer ! In any ol'
sawjints' mess yer'll get 'arf a dozen. Wot abaht Sawjint
Grice—gev Patsy Dunne a quid to lose the fight wi'
Nobby ? Wot abaht the sawjint-mijor—kep' a pub at
Barnes afore the war, an' 'ad two scullers ricin owver the
championship course for two 'undred a side, an' he findin'
both o' their stikes an' pyin' both men's trynin' expenses,
an' then gev 'em th' orfice w'ich man 'e'd 'ave win, as
soon as 'e'd got all 'is money put on ? Come to thet, wot
abaht this Sawjint Gort yer 'avin' all the jaw abaht ? "

" Well, wot abaht 'im ? " Tween's voice was bold, but
I guessed he had no sort of hope and was only calling
Pratt's bluff on the chance, and that chance an off one.
I looked at the Kid. The mask that he had kept all the
time on his face was all right, except for the colour, but
he was working his hands, unrolling and rolling up into a
ball a bit of white paper. I think he had hooked it out of
one of his pockets, to give his fingers something to do.

Pratt didn't mind going on. "'Ere's a bit abaht 'im," he said. "Ow, a tysty bit! Did 'is own side dahn. Remember 'ow the Koylies beat us, dahn at Bulford Kemp. One gowl to nuffink, an' we plyin' agen three parts of a gile o' wind orl the first 'arf, an' then the wind droppin' dahn dead soon as we'd crossed? Yuss?"

We all remembered that blow. "Lorst on the toss!" Pratt yelped triumphantly. "Lorst by a swizzle, an' Gort done it on us. 'E was ahr skipper. Ah think 'e might 'a been fytheful." Whenever Pratt tried to do the sob-story stunt I always thought his whimper was worse than his squeal. "'E torsed, an' the Koylies' skipper said 'Eads! It was a shillin'."

"Thot's reet," Barnard confirmed. "Ah was theer, plaayin'."

"There y'are!" Pratt's voice shrilled up into an argumentative treble. "A shillin'!"

Somehow we all felt that this identification made things rather black. I don't know why.

"T'shillin' fell on edge," Barnard further deposed, "an' rowled under taable, reet oot o' siight."

"Wot did I sye?" Pratt crowed over us all, as if this had made out his case. And I can't deny that we felt there was a lot of evidence knocking about, whatever it all came to The little vitriol-squirt went on with it, exulting. "Ol' man Gort dived under the tible, an' come up with the shillin'. ''Eads, all right,' 'e says. 'You got us.'"

"Wot ow!" Tween piped up, in joy. "'Ow's that for strite? Good ol' iron!"

124

THE FIRST BLOOD SWEEP

The Kid's face had begun to light too It was strange the small things that worked on that boy. His hands had stopped their quick fretting; they were unrolling slowly the white bit of paper.

"Strite!" Pratt squeaked. "You wite a mo. I says to 'im arter the match, 'Ya silly mug,' I says—'e was a private then—'w'y cawn't yer tike a chawnce when yer got it?' 'Wot chawnce, boy?' says 'e. 'W'y,' says I, 'at the torse. W'y didn't yer sye it was tiles?' 'It *was* tiles,' says Gort. I fair let a shriek: 'Gordelpus! W'y th'ell didn't yer sye so?' 'Because,' 'e says, narsty an' shawt, 'if I 'ad I might 'a been tiken for you.' 'E'll 'ave sol' thet match. 'Ullow! an' there 'e is agen!'"

We could hear the splash coming near in the trench, and the drip on the stairs. But we could only just hear, for the earth all round was fairly booming now with the muffled beat of drum-fire. The Kid leapt up and buckled his belt. The green was not out of his face yet, so quick was the change, but the living joy had come back. We had all tried to cheer him, and only made matters worse, and now that gutter-sparrow had tried to make matters worse, and had cheered him. "Now ye'll 'ear 'is yngel voice a-callin'," Pratt sneered.

Gort's voice did come. What a voice! It had always had a gallant clang, as I told you. To hear it would brace any nerve you had got. But now it was more than all that. Something had happened to it—what happens to wine when the sparkle gets in; it was bubbling with some kind of stir that had come to the man, though the words were nothing themselves—"Pay attention, in there!"

Pratt jibed, in a whisper, "''Ark at 'im, 'Anney. ''Anney, 'Anney, come aht and be slortered. Yer keepin' me aht o' me money.'"

Hanney laughed in Pratt's face, and then Gort blew the orders on that bugle voice that he had: "Barnard, on guard! Hanney, stand by!"

Hanney stand by? What the deuce——? "'Oollo! 'Oollo! Wot's this?" muttered Barnard. He was befogged. He did not care a curse that, while he went up into trouble, Hanney should stay, for the moment, in cover. But he had lost hold. His world was not working according to plan. Gort was not taking a chance when he got it. "Coomin', sergeant," he shouted, and then grumbled low again: "Wot gaame can 'e 'ave got on?"

"Noan!" Ince's voice rushed up like a rocket. "'E's streight!"

"Gwan, y' 'oly Juggins!" said Pratt. "Gort's sold the Kid. Thet'll be it. An' nah 'e's doin' a bit o' the dirty on 'oever's bought 'im—keepin' the Kid in aht o' the rine."

Hanney did not look at Pratt. "I'll stand by," he whispered to me, "in the trench."

"You will not," said I, being in charge. But he slipped me before I could grab him. The next I heard was Gort heading him back, with a jovial rant in his voice: "You're not paid to choose the time you'll be shot, my boy. Off with you, back!"

Tween chuckled at that. "Gowan, Bawnard," he taunted cheerfully. "'Op it, Bawnard! Aht of it! Gow aht, Bawnard, an' git put to deaf."

Barnard was dawdling to snarl; "Wojjerbet that——'

126

when one of the finest wild blasts of Gort's voice came blowing along to me : "Corporal Head, are you shifting that loafer ?" I hustled out Barnard, mumbling of bets and half-dollars, as Hanney came back through the smoke. He still held by a corner that old slip of paper. But he had forgotten it.

"Down't gow an' lose yer benk-nowt," said Tween, just to say something.

The Kid's mind was somewhere away. "What ?" he said. "This ?" When he saw what was meant he let the thing fall.

"Al'ays do to light a cig," said Tween, who was always short of a match. He picked up the paper and smoothed it, to fashion a spill. The next thing, he let a yell : "Chrahyst ! if 'e yn't drawn the sawjint ! "

Hanney stared at the noise. He did not take in, for a moment, that this was the slip he had drawn from a tin hat, as we all did, the morning he joined us, and had not even looked at.

Pratt helped him. Pratt's little dustbin of a mind was choke full of bits out of old movie plays, where everything works so that either one fellow has got to be killed or another. "You or 'im, 'Anney," he kept saying now. "It's you or 'im." Pratt was happy. A bit of real life seemed to be shaping at last like the trash that he enjoyed.

You could see the thing itself making its way in the mind of the Kid. You could almost feel his thoughts come, bit by bit—how he had been a beast ever to think of Gort's putting him up to be hit—and how Gort was probably

wangling it now, to keep him alive—and how **Gort** himself had perhaps been killed by this time, and the blood-
money gone not to Gort but to him, the Kid, who had
whined in his heart. It all worked in him, clear to see, like
the apple in some scraggy throat, till I jumped up and said,
" No, you don't ! " and just got a hold upon him before he
could make a break for the open again.

I was getting him tame when Gort's call of " Corporal
Head ! " came ringing again.

I shouted back, " Sergeant ? "

" Double out here," he called.

" Corporal Mason, take charge," I sang out, and I
doubled off through the smoke.

It was noon on a bleak clear day, clear without any sun,
and the light seemed naked and pale and aghast, like a
fainting face, after the warm, yellow gloom in Old Death.
All the raw chalk in the open looked chilly and haggard,
and wicked little whimpering winds were leaping about
as they do just before the cruel bald dawns that you get in
black frosts. When my tin hat went through our old hatch
of a door it was tinkling at once with the little bits of
metals and earth and stone and the like that kept falling.
The Germans were doing it well. All round us, above,
the level ground was jumping and splashing up everywhere,
just as a puddle does in a rainstorm.

But Gort's face took me most. It had changed more
than his voice. It had always seemed to me to be screwed
up a bit, as if he were holding it tight in some shape that
he thought was the best. It had gone easy now. He was
like someone well rid of all sorts of anxiety. " *Trop de*

tintamarre ! " he said, with the first smile I had ever seen on him. "*Trop de brouilliamini !* "

I had not known he knew Molière, or anything else except " Infantry Training, 1914." *That* he knew like " Our Father which art." Just a grand sergeant the man had seemed to me till now ; nothing else in the world.

" Not real drum-fire yet," he said, like a collector rejecting a piece that is not quite the thing. " A little *pas d'intimidation* only. Gad, but he's getting on to the Jocks, though. O the men ! The dear men ! " Where we stood was the right flank of our sector, so we looked down the trench where the Gordons were, on our right, and Gort had just seen a great shot with an enemy trench-mortar bomb.

It had plumped, at that moment, as fairly into the trench as a golf-ball prettily lofted over a stymie. There it had thrown up a great mound of chalk, damming the trench and rising a good yard above the parapet level. Out of the top of this mound a pair of legs, properly booted and putteed, were now sticking straight up into the air, from half way up the shin. I could have sworn they were just moving a little. Down in the mud of the trench, on our side of the dam, a kilted man, with his buttocks blown away and his body half-naked, was twisting like a cut worm and screaming for someone to put a bullet through his head. In an instant a little Jock private had rushed up on to the mound, in sight of enemy and friend, and begun to tear with his hands at the earth round the two legs, like a terrier when it gets frantic with trying to dig out a rat. In his frenzy of clawing his fingers looked long and hooked, like some great savage bird's. It was no good, of course. For a marvel he lived it out for a second or two hoisted up there like a target, and then a hulking Jock sergeant scrambled up to him and grabbed his collar and flung him down, like a kitten, into the shelter below, and dived after him, safe.

As they say, it put me in mind of my end. But there was work to do, luckily. "You'll take charge of the sector," said Gort. "I'm shirking till this little trouble is over."

The sergeant could say things like that. We knew him—at least, a bit of him—not all that I knew by now. For now I had found out the kind he was. I had suddenly slipped into knowing it, just as your heel will slip suddenly into a boot. I had seen a man like him before, when I was a boy—one that was always glum and sticky and not at his ease till a sailing-boat had sunk with three of us on her in winter, a good mile from shore. He was a wit and a happy man for an hour, until we got through. And then he had gone dull again. I suppose he was born about a wreck below par, and it took a pretty thick danger to give him the run of all the great stuff he had in him. The sergeant was like him.

"I've posted the sentries," he said to me.

"Bar one," said I, "sergeant."

"Oh, Hanney?" he said. "He can wait. I've a use for that post. Carry on, you, patrolling. And, for the love of God, don't let the young uns get their heads free." He was eager to get me away.

I went along to our left. It was a miracle. Not a man

130

had been hurt, above a small scratch. Gort was a wonder at keeping down losses. He used to coach and practise the new men in judging the enemy's trench-mortar bombs, like catches at long-on in cricket, and jumping round corners of trench to put solid earth between them and the burst—and also in not trying, ever, to judge the rifle-grenades, but just lying low till the thin hiss or spit of the flight had finished up in the nasty metallic rip of the little beast's bursting. Every man that we had was trench wise, thanks to Gort, bar the Kid, who had not had time to learn anything yet.

I came back by degrees to our right, stopping at each post to buck up the sentry there with some chat. When I came to the mouth of the little curved sap running out to the post where Hanney had been I thought I would give a look in. It was on the near lip of a big mine crater, the enemy holding the opposite lip, with a few yards of air between their rifles' muzzles and ours. Gort was there, stooping to peer through the steel crack of a loop-hole. His back was a sight—like a cat's when she quivers and wags her haunches with joy, crouching and watching a bird.

While I was reporting all well, a fat T.M. bomb started waddling across through the air, with its timber tail waving. "Resist not evil," Gort said, with his new happy smile, "with your head." We judged the ball, called "Left!" together, and jumped round an angle of earth in good time. It was near, though. I like them a bit farther off, but Gort's eyes were shining. They had a wild glee like a boy's when he hides in a bush, at some game, and the other side come close and don't see, but go on. Just to be chased and not caught, shot at and not hit—that's what one kind of man wants; it's his little nip of strong drink, and, wherever it's going, he'll never keep off it.

"Sergeant," I said, "do you know, this is very in-and-out running." Of course it was cheek, and the more so as he had been always, to me, of the you-be-damned sort that will blast your eyes out of your head at the start of a talk and not go back very much on it afterwards. Still, I felt it would go. He was clean off the earth, I could see, he had lifted right clear of all fear and fuss and the pride of his rank, and all the whole boiling of little mean things.

He cocked an eye at me lightly, not angry a bit. "Hullo," he said, "these be hard words."

"You know," I said, "what you drew in the sweep?"

"Sweep?" he said. "For the Derby?"

"The first-blooder," said I, "for this tour in trenches."

"God knows," he said; "I fancy I looked, at the time. Anyhow, I've forgotten."

He had. That was certain. "Well, sergeant," said I, "is it time I got Hanney?"

He gave a look round, with a flicking look on his face, like a dog scenting. "The muck's blowing over," he said. I could have sworn his face fell. I could not have said, myself, yet, that the firing was less. "Go the round again now," he went on, "and see are all fit."

"And send Hanney along?" I said quickly. I felt it was urgent—I couldn't tell why. It was as if something,

I couldn't tell what, had just begun to go on, and had to be stopped on the nail, like a fire.

He looked round and listened, again, sizing up the enemy's fire. He had been right about it, first time. A bombardment is just like a lot of notes played on a key-board with changes of pace. I could tell now, it was slow-ing. Gort seemed to think for a moment, not minding. " Sting in the wine of being," I think he said to himself ; " salt in the feast." And then to me, " Send him along."

I almost ran to the Trap. The storm was fast dropping into a calm. But you know how storms drop : with savage little returns, now and then, of their old pelting fury. Looking back as I went, I saw that nearly all of these last spurts seemed to fall near the crater. The dead calm was elsewhere.

I yelled to Hanney through the hatch, and he came bolting out like a whippet dog released to its master. Just at the moment he burst through the hatch I had a queer feeling that all the hurry was off. It was as if some-thing too tight had gone snap. The urgency of the business seemed suddenly gone. As we started along the trench a little haystack of smoke, with the flame just put out in its heart, was drifting away from the crater. The sound of the burst, long after all others, came like the last dropping bark a dog gives at the end of a great bout of temper. Not another thing fell.

I had my job to do, urgent or not. It did not take us long to reach the post on the lip of the crater. It was not much shattered. Rifle-grenades make no earthquakes like shells, but they can dissect pretty small. Nothing that you could have called a dead sergeant was left. There are walking cases and stretcher cases, and there is the ground-sheet case that only needs search and collection.

The boy, who had been in the leash half an hour, began to cry now and said he was shamed and had let the man die for him. But I bade him look to his front and I got the place cleaned.

The babble of voices there was in the Trap that evening!—and I with some sort of a letter to write to Gort's wife. Each talking after his kind, man or beast or creeping thing, as God had created him. Mason had to be saying, every few minutes, that it's an ill wind that blows nobody good. Barnard had drawn Tommy Tween in the sweep, and now he was sticking it out that Gort's death was not formally proved. " Is 'e identified ? " " Show me t'body." " Wot's a near forefoot to go by ? " When he was not saying one of these things he was saying another. Pratt was sucking up to the winner already : " Wot abaht it, 'Anney ? A little 'arf pint, 'long o' me, fust dye arter we're aht, jus' to wet the good luck ? "

" O yes, if you like," Hanney said, to be rid of him Hanney had been writing too. He came across to me presently with his letter. " Shall we send them together ? " he said. I guessed where his letter was for, but I wondered how he knew about mine.

IN HANGING GARDEN GULLY

TO climb up rocks is like all the rest of your life, only simpler and safer. In all the rest of your life, any work you may do, by way of a trade, is a taking of means to some end. That end may be good. We all hope it is. But who can be sure ? Misgiving is apt to steal in. Are you a doctor—is it your job to keep all the weak ones alive ? Then are you not spoiling the breed for the future ? Are you a parson or politician or some sort of public improver, always trying to fight evil down ? May you not then be making a muff every day of somebody else who ought to have had his dragon to fight, with his own bow and spear, when you rushed in to rob him and the other little St. Georges of discipline and of victory ? Anyhow, all the good ends seem a good long way off, and the ways to them dim. You may be old by the time you are there. The salt may have lost half its savour.

No such dangers or doubts perplex the climber on rocks. He deals, day by day, with the Ultimate Good, no doubt in small nips, but still authentic and not watered down. His senses thrill with delight to find that he is just the sum of his own simple powers. He lives on, from moment to moment, by early man's gleeful achievement of balance on one foot out of four. He hangs safe by a single hand that learnt its good grip in fifty thousand years of precarious dodging among forest boughs, with the hungry snakes looking up from the ground for a catch like the expectant fieldsmen in the slips. The next little ledge, the object of all human hope and desire, is only some twelve feet away—

about the length of the last leap of that naked bunch of
clenched and quivering muscles, from whom you descend,
at the wild horse that he had stalked through the grass.
Each time that you get up a hard pitch you have succeeded
in life. Besides, no one can say you have hurt him.

Care will come back in the end : the clouds return after
the rain ; but for those first heavenly minutes of sitting
secure and supreme at the top of Moss Ghyll or the Raven
Crag Gully you are Columbus when he saw land from the
rigging and Gibbon when he laid down his pen in the
garden house at Lausanne. It's good for you, too ; it makes
you more decent. No one, I firmly believe, could be
utterly mean on the very tip of the Weisshorn. I could,
if I had known the way, have written a lyric about these
agreeable truths as I sat by myself in the tiny inn at
Llyn Ogwen where Telford's great London-to-Holyhead
road climbs over a pass between three-thousand-foot
Carnedds and Glyders. I was a convalescent then, condemned still to a month of rest cure for body and mind.
But it was June, and fine weather. Rocks had lately
become dry and warm.

There are places in Britain where rock-climbing cannot
honestly be called a rest cure. I mean, for the body. Look
at the Coolin—all the way that a poor invalid must tramp
from Sligachan southward before he gets among the rough,
trusty, prehensile gabbro, the best of all God's stones.
Think of Scawfell Crag, the finest crag in the world, but its
base cut off from the inn by all that Sisyphean plod up the
heart-breaking lengths of Brown Tongue. From Ogwen
you only need walk half an hour, almost on the flat, and

then—there you are, at the foot of your climb. The more I considered the matter, the more distinctly could I perceive that my doctor, when saying "Avoid all violent exercise," meant that if ever I got such an opening as this for a little "steady six-furlong work," as it is called in the training reports, I ought to take care not to miss it.

But I was the only guest at the inn. And to climb alone is counted a sin against the spirit of the sport. All the early fathers of climbing held the practice heretical. Certainly some of them—Whymper, Tyndall, and others—climbed by themselves when they had a mind to. Thus did King David, on distinguished occasions, relax the general tensity of his virtue. But these exceptions could not obscure the general drift of the law and the prophets of mountaineering. Then came another pause-giving reflection. If, as the Greeks so delicately put it, anything incurable happens while you are climbing alone, your clay is exposed, defenceless and dumb, to nasty *obiter dicta* during the inquest. "Woe unto him," as Solomon says, "who is alone when he falleth!" Insensate rustic coroners and juries, well as they may understand that riding to hounds in a stone-wall country is one of the choicer forms of prudence, will prose and grumble over extinct mountaineers. Their favourite vein is the undesirable one of their brother, the First Clown in *Hamlet*, who thought it a shame that Ophelia (she seems to have slipped up while climbing a tree) "should have countenance in this world to drown or hang herself more than her even Christian."

No mean impediments these to a sensitive, conscientious nature's design for seeking health and joy among the attractive gullies and slabs that surround Llyn Idwal. Against them I marshalled all that I could remember of St. Paul's slighting observations on the law; also any agility that I had gained in the Oxford Greats school in resolving disagreeable discords into agreeable higher harmonies. Black was certainly not white. Still, as the good Hegelian said, black might, after all, be an aspect of white. In time it was duly clear to my mind that sin lies not in the corporal act, but in the thoughts of the sinner. So long as the heart sincerely conversed with the beauty of the truths on which rested the rule of never climbing alone it mattered little what the mere legs did : your soul was not in your legs. One of casuistry's brightest triumphs had been fairly won, my liberty gained, my intellectual integrity saved, my luncheon sandwiches ordered for eight in the morning—when somebody else arrived at the inn.

He stood confessed a botanist—he had the large green cylindrical can of the tribe, oval in section and hung by a strap from the shoulder, like the traditional *vivandière'* little cask in French art. He was also, I found while we smoked through that evening together, a good fellow. He had, too, a good leg, if one only. The other was stiff and unbendable at the knee. He had broken it last year, he said, and the bones seemed to have set only too hard, or else Nature had gracelessly grudged to the mended knee-joint of her lover a proper supply of whatever substitute she uses for ball bearings.

His name was Darwin. "No relation, really," he humbly assured me. His father was only some obscure squire. The son's Christian name had been Charles at the font, but on coming of age the dear fellow had felt it immodest to prey any more than he need upon his eponymous hero's thrice-honoured names. So he had meekly converted the Charles by deed poll into Thomas. This lowly and beautiful gesture convinced me, as you may suppose, that here was the man to go climbing with. He was indeed one of the innocent, one-thoughted kind that wake up happy each day and never turn crusty, and always think you are being too good to them.

One lure alone had drawn him to these outworks of Snowdon. Some eccentric flower grew on these heights, and a blank page in one of his books of squashed specimens ached for it. Was it so lovely ? I asked, like a goose. He was too gentle to snub me. But all that fellow's thoughts shone out through his face. Every flower that blew—to this effect did his soul mildly rebuke mine—was beauteous beyond Helen's eyes. All he said was : " No, not fair, perhaps, to outward view as many roses be ; but, just think !—it grows on no patch of ground in the world but these crags ! "

" It is not merely better dressed," said I, " than Solomon. It is wiser."

It was about then, I think, that the heart of the man who had gone mad on the green-stuff and that of the man who knew what was what, in the way of a recreation, rushed together like Paolo's and Francesca's. What had already become an *entente cordiale* ripened at tropical speed into alliance. Darwin had found a second, half-invalided perhaps, but still the holder of two unqualified legs, for to-morrow's quest of his own particular Grail.

To me it now seemed to be no accident that Darwin had come to the inn : it was ordained, like the more permanent union of marriage, for a remedy against sin, and to avoid climbing alone.

We got down to business at once. A charming gully, I told him, led right up to the big crag over Cwm Idwal. Not Twll Du, the ill-famed Devil's Kitchen. That, I frankly said, was justly *detestata matribus*—wet and rotten and lethal, and quite flowerless too. My gully, though close to that man-eating climb, was quite another affair. Mine was the place for town children to spend a happy day in the country : the very place also for starting the day's search for the object of Darwin's desire. In saying this, too, I was honest. Lots of plants grow in some gullies ; ferns, mosses, grasses, all sorts of greens flourish in a damp cleft, like hair in an armpit ; why not one kind of waste rabbit-food as well as another ? You see, I had not been a casuist merely, before Darwin came. I had used the eyes Heaven gave me, and reconnoitred the gully well from below, and if any flower knew how to tell good from bad, in the way of a scramble, it would be there. I ended upon a good note. The place's name, I said impressively, was Hanging Garden Gully, no doubt because of the rich indigenous flora.

His eyes shone at that, and we went straight to the kitchen to ask Mrs. Jones for the loan of a rope. I had none with me that journey : the sick are apt to relinquish improvidently these necessaries of a perfect life. Now, in the classics of mountaineering the right thing in such cases of improvised enterprise is that the landlady lends you her second-best clothes-line. Far happier we, Mrs. Jones having by her a 120-foot length of the right Alpine rope, with the red worsted thread in its middle. It had been left in her charge by a famous pillar of the Scottish Mountaineering Club till he should come that way again. " The gentleman," Mrs. Jones told us, " said I was always to let any climbing gentlemen use it." Heaven was palpably smiling upon our attempt.

The sun smiled benedictively, too, on the halt and the sick as they stood, about nine the next morning, roping up at the foot of their climb. " A fisherman's bend," I took care to explain, as I knotted one end of the rope round Darwin's chest.

" The botanical name," he replied—" did I tell you ?— is Lloydia." How some men do chatter when they are happy ! Can't carry their beans.

We were not likely to need the whole 120 feet of the rope. So I tied myself on at its middle and coiled the odd 60 feet round my shoulder. " A double overhand knot," I confessed, as I tightened it round me. " A bad knot, but for once it may do us no harm."

" The vernacular name," said the garrulous fellow, " is spiderwort."

" Tut, tut ! " I inwardly said.

The lower half of that gully was easier than it had looked : just enough in it to loosen your muscles and make you want more. Higher up, the gully grew shallow and had greater interest. The top part of all, as I remember it now, might be called either a chimney or crack, being both. In horizontal section, it was a large obtuse angle indented

into the face of the crag. The crag at this part, and the
gully's bed with it, rose at an angle of some 60 degrees.
Now, when you climb rock at an angle of 60 degrees the
angle seems to be just 90. In early mountaineering
records the pioneers often say, " Our situation was critical.
Above us the crag rose vertical," or, " To descend was
impossible now. But in front the rocky face, for some
time perpendicular, had now begun to overhang." If you
take a clinometer to the scenes of some of those liberal
estimates you blush for your kind. The slope of the
steepest—and easiest—ridge of the three by which the
Matterhorn is climbed is only 39 degrees. But this, though
not purely digressive, is partly so. All that strictly had
to be said was that an upright and very obtuse-angled
trough in smooth rock that rises at 60 degrees cannot be
climbed.

But in the very bed of our trough there had been
eroded, from top to bottom, a deepish irregular crack in the
rock. Into this crack, at most parts, you could stick a foot,
a knee, or an arm. Also, the sides of the large obtuse angle,
when you looked closely, were not utterly smooth. On the
right wall, as we looked up, certain small wrinkles, bunions,
and other minute but lovable diversities in the face of the
stone gave promise of useful points of resistance for any
right boot that might scrape about on the wall in the hope
of exerting auxiliary lateral pressure, while the left arm
and thigh, hard at work in the crack, wriggled you up by
a succession of caterpillarish squirms. This delectable
passage was 80 feet high, as I measured it with my ex-
perienced eye. An inexperienced measuring-tape might

have put it at fifty. To any new recruit to the cause—
above all, to one with a leg as inflexible as the stoniest
stone that it pressed—I felt that the place was likely to
offer all that he could wish in the line of baptisms of fire.
Still, as the pioneers said, to descend was impossible now :
the crack was too sweet to be left. And Darwin, thus far,
had come up like a lamplighter, really. I told him so,
frankly. Alpine guides are the men at psychology. Do
they not get the best out of the rawest new client, in any
hard place, by ceasing to hide the high estimate that they
have formed of his natural endowment for the sport ?
"*Vous êtes—je vous dis franchement, monsieur—un
chamois ! Un véritable chat de montagne !*"

I was leading the party. I was the old hand. Besides,
I could bend both my knees. Desiring Darwin to study my
movements, so that he presently might—so far as con-
formity would not cramp his natural talents—copy them
closely, I now addressed myself to the crack. When half-
way up I heard the voice of a good child enduring, with
effort, a painful call upon its patience. " Any Lloydia
yet ? " it wistfully said. Between my feet I saw Darwin
below. Well, he was certainly paying the rope out all
right, as I had enjoined ; but he did it " like them that
dream." His mind was not in it. All the time he was
peering hungrily over the slabby containing walls of the
gully, and now he just pawed one of them here and there
with a tentative foot—you know how a puppy, when first
it sees ice, paws the face of the pond. " These botanists ! "
I thought. " These fanatics ! " You know how during
a happy physical effort—a race or a hunt, a fight or a

game—you think, with a sort of internal quiet, about a lot of old things. There came back to my mind the old lines that I had once had to make Latin verse of :

> How vainly men themselves amaze
> To win the palm, the oak, or bays,
> And their incessant labours see
> Crowned from some single herb or tree.

Meanwhile I took a precaution. I first unroped myself. Then I passed the rope, from below, through the space behind a stone that was jammed fast in the crack. Then I roped myself on again, just at my old place on the rope. A plague of a job it was, too, with all those 60 feet of spare rope to uncoil and re-coil. But you see how it worked : I had now got the enthusiast moored. Between him and me the rope went through the eye of a needle, so I could go blithely on. I went. In the top of the crack I found a second jammed stone. It was bigger than number one : in fact, it blocked the way and made you clamber round outside it rather interestingly ; but it, too, had daylight showing through a hole behind it. Sounds from below were again improving my natural stock of prudence. You can't, I thought, be too safe. Once more I unroped, just under this chockstone, and pushed the rope up through the hole at its back. When the rope fell down to me, outwards over the top of the stone, I tied on again, just as before, and then scrambled up over the outer side of the stone with an ecstatic pull on both arms, and sat on its top in the heaven that big-game hunters know when they lie up against the slain tiger and smoke.

If you have bent up your mind to take in the details,

you will now have an imposing vision of the connections of Darwin and me with each other and with the Primary or Palæozoic rocks of Cambria. From Darwin, tied on to its end, the rope ran, as freely as a bootlace runs through the eyelets, behind the jammed stone 30 feet above his head, and then again behind my present throne of glory at the top; then it was tied on to me; and then there were 60 feet, half its length, left over to play with.

Clearly Darwin, not being a thread, or even a rope, could not come up the way that the rope did, through the two needle-eyes. Nor did I care, he being the thing that he was, to bid him untie and then to pull up his end of the rope through the eyes, drop it down to him clear through the air, and tell him to tie on again. He was, as the Irish say of the distraught, " fit to be tied," and not at all fit for the opposite. If he were loose he might at any moment espy that Circe of his in some place out of bounds. There seemed to be only one thing to do. I threw down the spare 60 feet of the rope, and told him first to tie himself on to its end, and then, but not before, to untie himself from the other. I could not quite see these orders obeyed. A bulge of rock came between him and my eyes, but I was explicit. " Remember that fisherman's bend ! " I shouted. Perhaps my voice was rather austere ; but who would not forgive a wise virgin for saying, a little dryly, to one of the foolish, " Well, use your spare can " ? As soon as he sang out " All right " I took a good haul on what was now the working half of the rope, to test his knot-making. Yes, he *was* all right. So I bade him come up, and he started. Whenever he looked up I saw that he had a wild, gadding eye; and whenever he stopped to breathe during the struggle he gasped, " I can't see it yet."

He came nearly half-way, and then he did see it. He had just reached the worst part. Oh, the Sirens know when to start singing! That flower of evil was far out of his reach, or of what his reach ought to have been. Some twelve feet away on his right it was rooted in some infinitesimal pocket of blown soil, a mere dirty thumb-nailful of clay. For a moment the lover eyed the beloved across one huge slab of steep stone with no real foothold or handhold upon it——only a few efflorescent minutiæ small as the bubukles and whelks and knobs on the nose of some fossil Bardolph. The whole wall of the gully just there was what any man who could climb would have written off as unclimbable. Passion, however, has her own standards, beyond the comprehension of the wise:

> His eye but saw that light of love,
> The only star it hailed above.

My lame Leander gave one whinny of desire. Then he left all and made for his Hero.

You know the way that a man, who has no idea how badly he bats, will sometimes go in and hit an unplayable bowler right out of the ground, simply because the batsman is too green to know that the bowler cannot be played. Perhaps that was the way. Or perhaps your sound climber, having his wits, may leave, at his boldest, a margin of safety, as engineers call it, so wide that a madman may cut quite a lot off its edge without coming surely to grief. Or was it only a joke of the gods among themselves over their wine ? Or can it be that the special arrangements known to be made for the safety of sailors, when in their cups, are extended at times to cover the case of collectors overcome by the strong waters of the acquisitive instinct ? Goodness knows ! Whatever the powers that helped him, this crippled man, who had never tried climbing before, went skating off to his right flank, across that impossible slant, on one foot and one stilt, making a fool of the science of mountaineering.

I vetoed, I imprecated, I grew Athanasian. All utterly useless. As soon could you whistle a dog back to heel when he fleets off on fire with some fresh amour. I could only brace myself, take a good hold of the rope in both hands, and be ready to play the wild salmon below as soon as he slipped and the line ran out tight. While I waited I saw, for the first time, another piquant detail of our case. Darwin, absorbed in his greed, had never untied the other end of the rope. So he was now tied on to both ends. The whole rope made a circle, a vicious circle. Our whole caravan was sewn on to the bony structure of Wales with two big stitches, one at each jammed stone.

You see how it would work. When Darwin should fall, as he must, and hang in the air from my hands, gravitation would swing him back into the centre of the chimney, straight below me, bashing him hard against the chimney's opposite wall. No doubt he would be stunned. I should never be able to hoist his dead weight through the air to my perch, so I should have to lower him to the foot of the chimney. That would just use up the full 60 feet of rope. It would run the two 60-foot halves of the rope so tight that I should never be able to undo the bad central knot that confined me. Could I but cut it when Darwin was lowered into provisional safety, and then climb down to see to him! No; I had lost my knife two days ago. I should be like a netted lion, with no mouse to bite through his cords : a Prometheus, bound to his rock.

But life spoils half her best crises. That wretch never slipped. He that by this time had no sort of right to his life came back as he went, treading on air, but now with that one bloom of the spiderwort in his mouth. Apologising for slowness, and panting with haste, he writhed up the crack till his head appeared over the chockstone beside me. Then he gave one cry of joy, surged up over the stone, purring with pleasure, and charged the steep slope of slippery grass above the precipice we had scaled. " You never told me ! " he cried; and then for the first time I noticed that up here the whole place was speckled with Lloydia. The next moment Darwin fell suddenly backwards, as if Lloyd himself or some demon gardener of his had planted a very straight one on the chin of the onrushing trespasser in his pleasaunce. You guess ? Yes. One of his two tethers, the one coming up from behind the lower jammed stone, had run out; it had pulled him up short as he leapt upon the full fruition of his desire.

He was easy to field as he rolled down the grass. But his tug on the rope had worked it well into some crevice between the lower jammed stone and the wall of the crack. We were anchored now, good and fast, to that stone, more than three fathoms below. What to do now ? Climb down and clear the jammed rope ? Leave that lame voluptuary rioting upon a precipice's edge ? Scarcely wise—would it have been ? Puzzled and angry, I cast away shame. I knew well that as Spartan troops had to come back with their shields or upon them, or else have trouble with their mothers, a climber who leaves his tackle behind in a retreat is likely to be a scorn and a hissing. Still, I cast away shame. Ours was no common case; no common ethics would meet it. I untied us both, and threw both ends of the rope down the chimney; then I let Darwin graze for a minute; then I drove him relentlessly up the steep grass to the top of the crag, and round by the easy walking way down.

As we passed down the valley below, I looked up. The whole length of our chimney was visibly draped with the pendent double length of that honest Scots mountaineer's rope. " I don't really know how to thank you enough," Darwin was babbling beside me, " for giving me such a day ! "

But I felt as if I were one of the villains in plays who compromise women of virtue and rank by stealing their fans and leaving them lying about in the rooms of bad bachelors. Much might be said for climbing alone, no matter what the authorities thought. A good time it would be, all to myself, when I came back to salvage that rope.

ALL FOR PEACE AND QUIET

AS the sister went out of the ward she paused to look back, with the knob of the door in her hand.

"Boys," she said, in the voice that made babes of us all, "five minutes to get into bed." We knew that five minutes, no more and no less, it would be. The door closed behind her, the little pat noise of it putting a kind of full stop to her words.

Of thirty wounded men in the ward, twenty-two had been up for the day. We were the blest. But bliss was precarious. If we were not good, the sister might keep us in bed in the morning. So we eagerly slipped off and folded our socks and red ties and blue tunics and slacks. The shirt did not have to come off. A shirt by day, it was a nighty by night—a good plan, I can tell you, when any delay may cost dear. In five minutes, dead, the door opened ; the sister looked down the long ward.

I lay next but one to the door, so I saw what she saw. There were twenty-nine faces duly laid on their pillows. Some seven looked dull and bed-weary. Twenty-one others—I throw myself in, for I felt like it too—looked shiny and young with the glee that you feel when the life in you has taken heart to go on with a will, after a check. A twenty-ninth face, in the bed on my right, was a model in wax, awaiting only some one final touch of rigidity and refinement. Into the thirtieth bed, at the dim far end of the ward, a vast bulk, in a white shirt less vast, flew through the air from afar with astonishing momentum and dived in fear and shame under the blankets.

The sister took it all in. A sergeant-major had been lost to the world when she turned out a girl. But not one to rule wholly by terror. She went first to the wax-faced man on my right, that always lay on his back with his eyes open, scrutinizing the ceiling. He turned his face a few degrees at the touch of her hand on his wrist and smiled a little. He had a great dignity then: the austere prestige of the dying, who are an esoteric patriciate, lifted above desire and fear and all quarrels.

The sister went on to the far end of the ward, where the meteoric giant had gone to earth under his blankets. His great shafts of limbs were convulsing them now, in his efforts to settle down for the night; the bed looked like a small linen bag with a large and terrified cat imprisoned in it and plunging. The plunging instantly stopped when the sister began tucking in the disarranged elements of the bed.

" Late ! A corporal late ! " In the stillness I heard her gently upbraid, while she reconstructed. Then a big bony face, like a knee, came to the bed's surface. The face looked sheepishly grateful for these so great mercies.

> We must perish if Thy rod
> Justly should requite us.

That was the visible feeling. No doubt Lance-Corporal Martin was vowing, within, that he would be first man in bed, from that night onward, for ever.

Her tour of inspection finished, the sister was passing the foot of my bed, on her way to the door, when another huge frame of a man, Lance-Corporal Toomey, a rug-headed Rufus, half rose from the bed on my left—the end bed on our side of the ward.

" Sister," he hailed her. She looked.

" Ye'd do right, sister," he said, with a clarion distinctness that seemed to me purposeful, " to be firm with that man beyant at th' end of the room, or ye'll not have a moment of peace." He paused for a second. Somehow I thought of an angler expecting a rise. He went on in tones of virtuous indignation. " Flyin' into his bed like a Toc Emma bomb ! " Another momentary pause. No rise. " Aye, or an eagle itself," he appended.

A low thunderous noise came from the far end of the ward. The first chaos of this mass of sound was just beginning to receive form in the words " Is it faultin' me that he is, the contrary toad ? " when the sister raised a reprehensive finger.

Speech, divinest work of man, relapsed into its raw material of booming sound. This sank by degrees. Lance-Corporal Martin was still out to be good.

The sister looked down at Toomey, reining him in. In him, too, mighty reservoirs of formless sound were straining against sluices. " Why don't you corporals screen one another," she said, with the voice of the dove and the serpent's wisdom, " like sergeants ? "

" If anny sergeant," Toomey impressively opened, " in all the wide world, had lived with the Martins of County Fermanagh these twenty years back, the way I've had to enjure—— "

" We won't tell any tales," the sister decided, and Toomey's conditional sentence never achieved its apodosis. " Good night, all," the sister said at the door. " Don't talk too long."

She gone, the ward, for a while, was all in a buzz. It is always the way at first, in night nurseries. Not for long, though. Sleep swiftly invaded one simple organism after another. Soon a few specialists only were riding their hobbies on into the night. In the bed opposite to me the ward grouser was manfully keeping it up. " Call them ole dishwashin's supper ! Well, roll on, breakfus' ! 's all I 'ave to say." Three beds away on my right—two beds beyond the moribund man—the ward braggart, like Alexander the Great at the feast, was still routing his foes. " An Austrylian 'e was, an' 'e come powkin' 'is conk in me fice, aht in Ahmenteers. ' More'n twelve thahsand mile ah've come,' 'e says, ' to sive you ticks,' 'e says. ' Well,' ah says, ' you down't come a bloody inch further,' ah says, an' knocks 'im dahn. That's wot ah done. Fair on 'is boko. Knocked 'im silly, ah did. 'Struth ! "

To these shrill melodies a fortifying bass was furnished by my neighbour Toomey. Cowed, but not changed in heart, by the sister, he still rumbled on in his bed like a distant bombardment. Like Martin's last share in debate, it was mainly a deep rolling or heaving sound, not chopped up into words ; in its vague impressiveness it affected my ear somewhat as a distant range of undulating hills affects the eye. But now and then a short passage of firm, articulate commination lifted itself into clearness, as tiny islands of coral, emergent morsels of vast unexplored structures, stick out into daylight. " No peace an' quiet at all to be had ! " " Lettin' on to be wounded, an' he leppin' the width of the world into his bed ! " And I used to think once that soliloquies in plays were unnatural !

" He an' his Bloody Hand of Ulster! Let him put it in his pockut, th' owld miser! " " Boundin' into his bed wid a thraject'ry on him the like of a grasshopper! "

Toomey always seemed to return to that vision of Martin's retirement for the night as if it were fuel wherewith to stoke important fires. I cannot think that Martin's jump was intrinsically offensive to Toomey. All jumping, by man or beast, is highly valued in Ireland. Was it that any salient gesture of a foe's helped to make the whole abhorred image of him more provocatively vivid? I leave it to you; I am not a psychologist, only a reporter.

" The both'ration I've had wi' the Martins, the whole of the pack! " " If Our Lady of Seven Sorrows cud set eyes upon them she'd know what it is to be vexed! " You have heard the low seismic swallowings, pantings, and gulps of a big locomotive held in leash at a station. Toomey, making a good recovery from gunshot wounds and wrestling in spirit with a friend, was rather like that—a volcano tied down with string. Then reverie fell back on the old refrain. " An' he descindin' into his bed the match for an airiplane landin'! " " The spit of some owld hivvinly body bruk loose an' traversin' a very big share of the firmamint! " The obsession of Toomey by this haunting spectacle was becoming ungovernable. He had to communicate it.

" Did y'ever see the like of it—ever? " he said across to my bed.

" No," said I, feeling my diction somewhat jejune where so much picturesque figuration was going.

" 'Faith, then, *I* have," said Toomey, " the time that

I seen his mother's starved hens flyin' in at me own mother's door, the vulchures, fit to grab all the nourishment out of me plate, an' I a young child."

"Bit of a war on," I asked, "over there?" I wanted to sleep. I am only an Englishman. I do not hold two turbulent quarts of tireless life in the modest pint pot of my body. It seemed the shorter way to sleep—just to open the dykes and let the whole narrative flood escape and be done with, rather than lie there all night being splashed with these interjectional spurts from the dammed seething waters. "Bit of a war on?" I asked.

"Ye'd not know," he impressively said, "what war is till ye'd visit the County Fermanagh. Talk of this little unpleasantness wi' the Germans! Ach, nothin' at all! Here wan day an', mebbe, gone in the mornin'! Th 'owld wars of Troy'd be hard set to match what we have in Dromore, where the Martins are nex'-door neighbours to us, an' not respected by anny except black Prod'stants the like of themselves. To crown all, Jawn Martin beyant had no sooner grown up than nothin'd do him but join the police. Mind, there's manny a decent lad in the force, the way ye'd pity them on a summer night in Dromore— we in the licensed prem'ses, all at our ease, an' they walkin' this way an' that in the street, wi' the tongues hangin' out of their mouths, till the hour'd come they'd be free to lep in at the door and tell the boys they must shout or come out of it. Martin beyant would not give ye the choice. He'd liefer lose drink than not cause human mis'ry. A narra-spirited man an' a bigoted peeler!

"I thanked God for this tiff wi' the Germans. Ye see,

in a sense, it gev Martin, bound and gagged, into me hand. 'Jawn,' I said—I was just after enlistin'—'Jawn, I'll be back before long. I'm only steppin' over to France, to destroy,' says I, 'the greatest Prod'stant Pow'r in Europe.'

" ' I might be at him before ye,' says he, pensive-like; an' that made me unaisy. Ye'll understand, I'd always taken me own part the best way I could in this chat that we had, wan day an' the next, an' yet I'd the wish in me mind to have peace an' quiet awhile, away out of Dromore, leavin' that plague of the world confined where he was. An' so I'd thanked God for th' affair wi' the Germans. But now I was not aisy at all.

" ' Ye wouldn't sneak off,' says I, 'out of the country, an' leave all the poor German bands in the street unprotected from popular voylence—an' they nivver safe in Irelan' yet since eighteen and siventy?' If I'd had more diplom'cy, God help me! I wouldn't have said it. But I had me quarrel to mind. I've paid for it since.

" Says Jawn, ' There's a sowl of good, as they say, in annything evil. I see be the papers these Germans are smashing every Romish cathedral they'd meet on the road.' Half thinkin' it out, he seemed to be, for an' against, an' half stickin' the knife into me bowels. The same as meself. ' Don't fear to speak,' says I, ' of '98, an' of all that the Hessians done for ye then.' Madness it was to say that. I was drivin' him into the army, that might have been a haven of rest for meself for the whole of the war with him out of it. Leavin' all was I, an' baffl'n' meself, to have the wan dig at him.

" Jawn looked at me quarely. ' An' so ye'll fight for

th' English,' he says, ' afther all ? Ye're penitent, are
ye ? '

" However, I knew me facts there.

" ' Th' English are givin' free transport,' says I, ' to
this Wipers beyant, to see th' owld flag hangin' up in a
church that th' French and th' Irish tuk off them in Mal-
bury's wars. There's the fine penitence for ye,' says I.

" ' Oh, ye'll see Waterloo yet, me fine touris',' says he,
jus' to howld me in play, an' he puzzlin' it out what to do.

" ' I may that,' says I, ' if it's annyways near Fontenoy.'

" ' Dad ! ' says he irrel'vantly, scratchin' the back of
his head in the ag'ny an' sweat of his thinkin' about somethin' else, ' haven't I got a better right not to be friends wi'
the English than you, an' they just after passin' Home
Rule ? '

" ' What way'd ye be friends,' says I in me anger an'
folly, ' wid English or Irish, an' you after lookin' away
with the whole force of your eyes while th' Orangeys
landed their Mausers from Hamburg, to slaughter the wan
as lief as th' other ? '

" We bruk away then, he to-Hellin' the reel Pope, an'
I doin' the like be every weeny black Pope in his wrinkly
tin Vaticun all over Ulster. It was a kind of salute we had
always, meetin' an' partin'.

" Me word, if the nex' Monday mornin' he didn't come
shankin' on to parade on the wan barrack-square with
meself ! We were formin' B Comp'ny that day, an' th'
owld sergeant-major, Yorke, was expoundin' : ' Sixteen
men I'll have now, for Number Wan Section. Who-iver
they are, they'll work together, eat together, sleep

together, an' fight together, from this out to th' end of the war. So anny man wishful to be with his pal, let him look to it now, or howld his peace after. Men off'rin' for Number Wan Section, two paces step forrud.'

"Jawn gev me wan vinegar glance an' stepped out. 'Good!' thinks I, 'the sky's clearin' wance more. Number Sixteen Section's the boys for me money.' I sted where I was.

"A big man on me right was gazin' melancholy at Jawn standin' out there be himself. 'Be the grace of Gawd,' says the big man under his breath, 'he may not be lousy,' an' then he steps forrud.

"Another tall man on me lef' flank was a big grazier from Antrim. 'The two of them's evenly fed,' he says, soft an' low, an' he steps forrud too.

"'Three hefty merchants,' says Yorke. 'Anny more welter weights? Come, men; th' enemy's waitin'.'

"Ye'll notice big men herd together. The lads of six foot an' a bit began steppin' out in their pride.

"'Anny more Tiny Tims?' says Yorke, sizin' up wid his eye the ten giants that he'd collected be now. A good man he was, an' had his joke always. Killed in attackin' the Railway Triangle, east of Arras, an' the las' thing he said was, 'Get on wid ye, men! I'm makin' a separate peace.'

"Be now he had fifteen ver'table monsters. 'Wan more Little Tich is required,' he says; 'gran' feather-weight boxin' we'll have in this section.'

"You'll think me an ijjut. Me that had the chance to be shut of him then, for the duration, be the whole len'th

158

of a comp'ny ! I dunno. Was it the wicked conceit of me stachure ? Was it the shame of me bein' left there stickin' out of the ranks like a lamp-pos' surrounded be childer ? Was it that word of the boxin', an' all the good it might bring ? Dear knows. We're the quare creatures. I stepped out, be some divvil's guidance.

" I'll not trouble ye with a report on our trainin'. Ye've had it yourself. It was six months for all, an' six years for th' unfort'nates condemned to the wan hut wi' Jawn. We came out to the trenches above be the Brickstacks, opposite Quinshy, a good place in itself, wi' the makin's in it of quiet an' peace—the dug-outs more Christianable than manny ; some derelic' trucks on a line pretty handy with coal ye cud pinch ; our quartermaster-sergeant had genius—he gev us the first plum-duff that ever was et be British troops in a firin' trench. Nothin' to trouble at all, only Jawn's clapper tongue, would deafen a miller, murd'rin' the life of the home.

" Wan night he beat all. He went beyant the beyants. In the depths of the perishin' spring it was that they have in those parts, an' a nasty thin evenin'—mist at the time an' a cowld rain, an' the moon not risin' over La Bassy till later—a night the monkeys ye'd see in a Zoo would be huggin' each other to keep off the chill. An' Jawn there in the dug-out, settin' a little apart, hottin' his nose wid a short pipe ye'd be sick an' tired of seein' before ye all day, an' all th' off-duty men around him good practical Cath'lics, an' he just gabbin' an' gabbin' on about the wil' doin's of some wicked medjeeval Pope he'd invented.

" Of course I was settin' him right. Why wuddent I,

knowin' me facts ? But not enjoyin' it rightly ? Weary
I was of it all. I'd liefest have had the man dead, an' all
the worl' tranquil.

"Near direct action we were when Shane, our platoon
sergeant then——he was killed at Dickybush after——came
into the homestead. 'Gawd save,' he says, 'all in this
place. Two volunteers to go wirin' ! '

"We had been out for a while, so there wasn' th'ugly
rush that there was in the wild early days at anny word of
this kind——every man for some sport for himself before
they'd have a peace signed on him. Everywan knew be
now he'd be killed in his turn, widout leppin' out of his
place in the queue. Everywan knew it except a few lads jus'
come out on a draft. I had to spake quickly be reason of
them. But I'd have given me sowl for an hour's quiet that
minute, out of range of that Orangeman's endless haran-
guin'. An' so I let a yell that won be a head : 'Ye
promised me, sergeant, the time that I misbehaved in the
lorry.'

"The same instiant minute Shane nodded at me an'
Martin's owld creak of a voice came in a good second.
'Here's wan, sergeant,' he said. The boys from the draft
also ran. A Jock that was wan of them had a very good
argyment nearly ready why he should go, but they hadn't
an old soljer's rapid rayaction under a schimulus. So I
stepped down before them into the pool. An' they hadn't
enjured gen'rations of livin' nex' door to the Martins.
An' now, Gawd help me ! Martin was in for it too, an'
me quiet day in the country rooned before it had started.

"Dad, if he hadn' been an' actin' lance-corp'ral,

160

unpaid, I'd have been in command of the party, me bein' th' oldest soljer be four days. Then I'd have put him through it, be cripes. As it was, I had only me tongue to come to me succour.

' "Why was it,' said I, when Shane had tuk leave of us, ' why was it, corp., that I got in before ye? Aye, in a canter."

" ' Ye're a liar,' says he, with his teejus habit of missin' the point of annythin' that a person would say. ' I was as wishful to go as yourself.'

" ' Ye were that,' says I. ' Your only trouble was, when the time came for a good burst of speed, ye were thinkin' in English, a foreign language insthinctively quare to your mind, an' you a son of the Gael, however unworthy.'

" ' An how long would ye be, your own self, ye fantastic owld antiquarium,' says Martin, ' in gobbin' out even as much as " Here, sergeant " itself in the Gaelic? From this out till to-morra, mebbe, an' then needin' a pencil an' slate brought to ye before ye'd compose it.'

" Very fortunate was it that I knew me fac's. ' The mere gabbin',' says I, ' is nothin' at all. It's the thinkin' that matters. Learn your own native language, that's hangin' about at the back of your mind, an' think in it bowldly. There's no other way ye'll escape from this habit of strikin' in late in th' hour of trial—the way a hen would be in, an' she attemptin' to neigh or to bark it all out in her mind, instead of cluckin', before she'd quit out of the way of a cart. She'd be swif'ly run over.'

" He didn' answer me then.

" At eleven that night we p'raded for orders, relatin' to th' enemy's wire. The captain himself—he was killed at High Wood—was out an' about in the trench when we quitted for Germ'ny. Be that we knew we'd be safe from behin'. The *m'ral* of the men was that high they'd be loosin' off all night at anny owld clod they'd perceive, in the hope it might be a good action. As for annythin' movin', Gawd help it! An' so we were glad. If annythin' human could hol' back our sentries from makin' good groups on our sterns, Shane and the cap. were the men.

" Th' enemy's wire was seventy paces away, an', as you know, seventy paces route-march on your belly, the way of the snake in the garden, is no good to annywan, ever. An' we went about an' about. An' not a good motorin' road. An' Martin never stopped but wance to take breath. Mebbe ye've observed, when takin' a walk with a friend, there are men that go faster an' faster the more they're intent, in their secret insides, on inventin' the divvil's own repartees, all ready to fire, an' when they're gettin' on grandly within, forgin' all the munitions they'd want, they're fit to gallop itself. Jawn was wan of that breed.

" ' Th' only time he tuk ever a pull, he turned on me sudden an' ven'mous.

" ' Ye an' yer preestoric owld Gaelic,' he whispers. ' A match for th' owld women that won't use a ha'porth of linen that's made with anny machine that was new since the Flood!'

" ' For a man,' says I, ' tryin' to marshal his thoughts in an outlandish tongue, ye've not dawdled at all,' says I, ' in devisin' that smashin' retort. Very barely four hours.'

" ' An' why stop at the Gaelic ? ' he says, pantin'
' Why not go back to the reel startin' gate while ye're at
it, an' think in Persian or Greek or some owld yappin'
noise your forefawthers made, an' they monkeys leppin'
about in the trees ? '

" I was beginnin' at ' What are the fac's ? ' when
Martin adjourned th' altercaysh'n be boltin' again. It had
done no good to our tempers—to Martin's, at all events.
Well he knew I'd have put it across him on that new
quistion he'd raised, before settin' off, about sci'nce an'
religion. So nothin' would do him but plunge on ahead—
I'll engage he was sharp'nin' already the nex' dart he'd
plant in me vitals—an' not mind at all where he went till
he'd slung himself head over heels into wan of th' owld
flooded bog-holes left be the shells. It was cammiflaged
well, I'll admit, be the grass an' weeds growin' out long
from th' edges.

" I made sure th' en'my would hear. How wouldn't
he hear a rhinoc'rus, the like of that, ent'rin' his bath ?
An', the nex' thing, I knew it. Aye, be dint of a weeny
clickin' noise from his trench. ' It's the cock of a Verey
pistol,' thinks I, ' an' the lights' comin' on—that or some
godless breed of an Emma G that they have,' an' it bein'
called to attention before it'd start the good work. An' be
now the mist was liftin' like smoke, and a clear half-moon
was drawin' itself away up the sky, informin' upon us.

" Jawn's head was stickin' out of the deep. I'll engage
he was cold. But his body was sheltered from solids at
anny rate, under th' eastern wall of the hole, an' he could
submerge for short inthervals, too, if the en'my should

menace his per'scope. I was the victim, out in the wide world. An' Jawn saw it. ' Flatten yourself,' he whispers, over the face of the waters. ' Flatter be far, for the love of Gawd. Rowl yourself out into pasthry before they put the light on ye.'

" As if I hadn't been as flat as a plaice from that diaboluc click onward! Breadth had I been from that out—no thickness at all. I hadn't stopped to tilliphone to the Corps. An' when I was plast'rin the ground the like of a poultice I found a hole the size of a thimble an' put me nose in it.

" Then came the trouble. Th' en'my was damnin' expense. Pourin' out money. Three, four, five Verey lights at wan instiant. Wan great beast of a moon of an arc light that'd illum'nate half Waterloo Station went on loitherin' over me head ; I'd enjured a twelvemonth, be what I could judge, of invijjus public'ty before it went out. An' me there thinkin' slowly an' caref'lly, the way ye'll remember ye do at these crit'cal times. Thinkin' of all the blisth'rin' force of the chat I'd be givin' to Martin after lights-out. I that had come to that remote place for a season of quiet, away from the janglin' an' strafes of the world !

" In due course the fireworks expired. It was amazin' —the bats hadn't seen me at all. I'll own that the firs' thing I done with me new start in life was to spend a few seconds in grinnin' an' shiv'rin' with chuckles, alone with meself in the kin', friendly dark, the way I'd not done since the time that I was a boy an' successfully hid in the loft from me fawther.

" Me nex' thought was of others. Jawn had crep' forth

from the bed of the hole an' lay on the shelvin' foreshore like a newt.

"'Shall we shank on wid us, corp'ral?' I whispered. 'Or have ye come out,' I added, 'to this rugged spot to enjoy adult baptism only?' I knew me fac's, an' the pestilint sect of the man.

"'The game is quared on us,' he mutters discontentedly —more to himself than to annywan.

"'It's developed,' says I, 'be what I can see. Aye, into ruthless subm'rine warfare.'

"'Mebbe I'd do right,' says he, 'to go back for fresh orders. Th' en'my'll watch an' pray from this out.'

"'That will he,' says I—'the prayer of all sentries on earth—to rest unmolisted, them an' their wire, to th' end of their watch. An' shall we,' says I, 'grant their request? Isn't it carryin' gratichude altogether too far for the Reformaysh'n itself an' the lunch that the Kaiser gev Carson?' I'd have said more, but you know the neciss'ty of keepin' concise in the field.

"In the moon comin' up I could see the whites of his eyes very quare-shaped, an' vicious at that, the like of a horse that'd offer to bite ye. 'Carry on workin',' he says, an' he was off on the top gear agen, lettin' her rip for th' en'my's wire, the way I was out of me breath, beltin' after him on me intistines, before he let up all of a sudden. An' then I saw why.

"He'd his head lifted up from the ground an' pressed back, same as 'Heads backward—bend' at phys'cal drill, only frozen. Starin' rijjid he was, to his front—ye could tell be the back of his neck, an' it dark night, how he had the eyes bulgin' out of him, starin' at somethin' ahead. And then I saw it too.

"It was a man, lyin' full on his belly, the like of ourselves. He was facin' at us, an' he had his head a weeny shade lower than Martin's—a better posish'n. Crouchin' outside th' en'my's wire he was, ten feet away from us.

"Wan thing I knew. Jawn would do right. A quar'lsome man an' a bigot, an' yet a good soljer, the way he'd do best the time all before him'd be goin' its worst. He'd earned his stripe well. Put him in anny old sort of fix without notice—he'd do right, the like of a cat knockin' into a dog an' she walking about in her sleep—she'd have th' eyes of him out before lettin' a yawn, an' Jawn was her aqual.

"There was clouds now stravadin' over the moon, so I couldn't see all. But me heart told me Jawn was sof'ly extractin' a bomb from his right tunic-pocket, an' pickin' the pin of it out an' howldin' it ready to throw. An' so I conformed to his movements.

"He didn' throw yet. An' then he set off crawlin' precautiously on towards the man, an' the man keepin' still, the two of them there like a dog reconnoitrin' a strange dog, an' movin' gradjal an' stiff an' collected for action. Jawn crep' an' he crep' till he'd crep' right up to the man, an' then he put out a hand an' felt the face of the man, an' the man didn' mind it at all. And then Jawn signalled me on.

"Ye'll have guessed how it was. Fritz had been dead a great while. He'll have gone on some little errand or other, the same as ourselves. There'd been no battle there, at the time, an' dead men in the open not plentiful yet.

"I got me head level to Jawn's. Grand cover we had, behin' the dead German. 'How soon did you know?' I whispered.

"'Know what?' says he, sourly.

"'That Fritz,' says I, 'had gone west?'

"'From the moment,' says he, 'that he came in me presence!'

"'How'd ye see it?' says I.

"'I didn',' says he, 'I smelt it. The man was a Papist. Wouldn't I know a dead Papist at ten feet away in the good scentin' weather we're havin'?'

"Thinks I, 'Ye can score yourself wan, me fine lad, but the night isn't over.' I left it at that for the moment. 'An' so,' says I, 'the pin is not out of your bomb?'

"'It is not,' says he. 'Why'd I stir it at all, an' it apt not to go back to juty when wance demob'lized, an' have me there holdin' a mad trigger down for the rest of th' excursion?

"'Ye're lucky,' says I. 'Me own pin's out an' lost on me wholly, beyant in the grass.'

"'Hand me over the bomb,' says Martin.

"I may have hesitated a second.

"'Hand it across,' he repeated. 'We're not holdin' the first session after Home Rule.'

"I handed it over. Strainin' it seemed to be, under me hand, to be off an' at work in its station in life.

"'I'll give it a home,' he says, 'now, and a chance in life afther.'

167

"How Jawn kep' it still in his pocket I will not con-jicture—whether he made a new pin of the wire he'd always have somewhere about him for snarin' a hare, or tuk out the detonator itself, or what all. Annyway, we went on wid our own work, an' a long hour it tuk, th' en'my not bein' aisy about it at all. Constantly sendin' up flares an' bringin' about unemployment. We'd have had little done if the cap. hadn' worked the lights for us the grand way he did, makin' it black where we were be the force of contrast. We saw his hand in that an' thanked Hivven for a good off'cer.

"At las' we were through with our job on the wire, an' then we assimbled behind a good sizable mole-hill. We tuk the bearin's of home. Then Martin, that hadn' seemed to be troubled till then, says low in me ear, 'A nice way I'm in with this ingine of death that ye gev me rampin' roun' in me pockut.'

"Says I, 'Apt we'll both of us be to be killed be a primmature burst before we'd see home.'

"'We'd do right,' says he, 'to be quit of it early!'

"'Ay, an' wan or two more,' says I, 'be way of support. It's not twenty paces,' I says, 'to th' en'my's trench.'

"'An aisy throw,' he says, 'an' we'd be protectin' our-selves.'

"'Aye, an' wearin' down also,' says I, 'th' en'my's man power.'

"'They're affordin' the best mark,' he says, 'that we've had in the war. They're as thick in that trench as a p'lit'cal meetin'. Lay your ear to the groun' and———'

" ' Have I ever laid it annywhere else,' says I, ' durin' the whole of this outin' ? '

" He had an offensive way of goin' on from the las' word he'd said, f'r all the world as though you'd not spoken at all—' Ye'll hear,' he says, ' the deep hummin' sound of scores of guthral voices.'

" I was incinsed be that trick that he had of scornin' me little irrel'vance.

" ' Guthral, indeed ! ' says I. ' Worse than the English, bedad ! A good slam of the Gaelic I'll want, to clean me ears after it.'

" ' Are ye faultin' th' English again,' says he, ' an' you helpin' them now ? '

" ' Aye, to come to their senses ! ' says I. ' They that set up the Prooshans in life, at th' expinse of the French, when the Honezolluns hadn't the stren'th of the Great Joyces in Galway. They that went to the Germans for fiddlers an' waiters and kings, an' a cap for their troops, an' a new religion itself, an' whatever they'd want. An' why not at all, an' they all cousins an' Chewtons together ? '

" 'Aye, an' fightin' like cousins,' says he. ' Keep your head down,' he says. But he didn' say ' Keep your mouth shut.' A quar'lsome man, wild to be at an argyment always. An' there we lay, he on his right cheek and me on me left, nuzzlin' the wet ground while I told him the facts, subjuing me voice because of th' en'my. The two front lines were sendin' up lights all along, an' they intersictin' wan another above us, the way we seemed to be lyin' out on the floor of a church with a great arch over our heads an' the lines of it all drawn be the thracks of the lights.

" ' I give th' English all credit,' says I, ' for comin' at length to themselves, the same as Friday, the black, that turned friendly at last an' assisted the traveller, Robinson Crusoe, against the man-eating scuts that he had in his fam'ly.'

" ' Downright fawnin' ye are on th' Englishry now,' says Martin. ' Ye'd do right to get a good Blighty, the first wan that offers, an' off with ye home, an' preach your new gospel to haythens like Casement.'

" ' Such bein's are more than half English,' says I, ' be descint, or they'd see the gran' laugh we have now at th' English, an' they comin' roun' at last to detist all that's German as much as ourselves. It's th' English are well entered now to the vermin they had for their cousins, an', please God ! they'll never return to their vommut.'

" I could have gone on a great while, bein' up in me fac's, but ye'll understand I was speakin' in that place under some disadvantage. Still, I'd got me word in, an' I felt aisier after. But, if you'll believe me, he wasn't convinced, not a tittle. He took out the bomb from his pockut. ' Before ye have me desthroyed,' he says, ' screechin' an' hullabalooin' across to your fellow-en'mies of Englan',' we'll bring an event into those quiet lives.'

" Wishful I was for it too, after the teejum we'd had, and yet frikened to think of Shane and the cap., that had ordered us not to draw fire. ' Have we author'ty for it ? ' says I.

" ' Gawd help ye ! Author'ty ! ' says he. ' If ye hadn't turned the Reformaysh'n away from the door ye'd have had some indivijjal judgment to-day and a modern soljer's inish'tive in an emergency.'

" 'Reformaysh'n be damned !' says I. 'We'll have the cap. up on his hind legs, tellin' us off for self-actin' mules.'

" 'Ye're a child of author'ty,' says he. 'So strike out for the comforts of home, the while I'll be leavin' me card on our friends an' then comin' after you.'

It was an order. I set out and squirmed a good four paces westward. Then I thought : hadn' I a good right to disobey Jawn, an' he disobeyin' a comp'ny commander ? So I checked in me course an 'tuk the pin out of the second bomb that I had, an' lay on me back, with me feet to the foe an' me head illevated, to see what Jawn could be at. Black agin the full shine of the moon I saw his long arm, an' it swingin' back for the throw. Then I loosed me own trigger an' held on for two seconds good, to make sure of a burst before Fritz'd return the ball to the bowler.

" Jawn was a natty bomber, no quistion. His cracker had burst fair in th' en'my's trench be the time I had sent mine speedin' after its collaigue. The cries that there were in that trench ! An' then Martin came west wrigglin' headlong, the back of him ripplin' the like of a caterpillar tryin' to gallop.

" 'Canter,' says he, 'canter along on your bowels.' There was a great seren'ty, for wance, in his voice. An' I was the same way meself—at peace with mankin'. Ye know how ye are, after prosp'rin' at hom'cide. 'Canter,' says he, almost civ'lly. I cantered. An' reason enough. The entire concern was lookin' apt to degen'rate into a war of attrish'n. Fritz was dead tired of havin' the night that he'd had. Dead set agin anny more secret diplom'cy. The hivvens above and th' earth beneath, he illum'nated them all. 'Twas as though he'd tuk Paris. If he'd have done it in London he'd have been fined.

"All the way Martin behin' me was gruntin' out steerin' directions. 'Half-right,' he'd never done sayin', ' ye owld Maryolather—anny patch of dead ground in this wicked world is half-right.' Or 'Gallop,' he'd say, an' we close on our wire, 'nivver mind yer owld vitals. Gallop before they put the lantern on our posteriors.'

"Then it came on us. Aye, like the judgment of God fallin' down. There was but the wan lane through the wire, an' straight an' narrow the way, an' that fi'ry sword flamin' down on the gate. No use shammin' dead, like the beetles ye'd bring into the light with your spade. We'd ha' been filled as full of holes as a net. Fritz had got the addhress, an' already the stuff was not bein' far mis-delivered.

"I lep in at the op'nin' an' on towards the par'pet. 'Rowl yourself over it! Bowl yourself over it, anny old way,' yells Martin behin' me, an' then I knew he was hit, be the traces of voylent effort there were in the voice of him.

"Now, I'll engage ye'll think me a fool—I that had got, be the act of God an' the King's en'mies, a chance to be shut of him wholly, for ever, relievin' th' entire platoon of the curse of Hivven that Martin was never done callin' down be his godlessness on us all. A man's a quare thing. I'm told be a firs'-class dentist beyant at Dromore, if a tooth hasn't got a tooth the match of itself for to bite on it's apt to grow weaker an' weaker, an' fall out eventsh'ly. Mebbe it's like that. An' Jawn was as bad. ' Off wid ye to hell ! ' he says, houndin' me on into safety, th' instiant I halted. Mebbe it was just me resentment at him givin' orders an' curses an' business as usu'l wid all the pride of a corp'ral expirin' on juty. ' Be damned,' thinks I, ' if ye're goin' to stay out there actin' hayroes an' martyrs all night.' So I went back an' gev him a fireman's lift an' away with me lovely burden. He was the weight of the world. Be cripes, it's a good horse wins the Grand Nashnal!

"All the damage I felt on the road was a sting in me unemployed arrum. Then I took wan flyin' lep, the like of his own way of goin' to bed, an' the two of us landed, in wan knot of arrums an' legs, into the trench, alightin' first on the firin' step an' then in th' stable-washin's of water there was in the fairway below, an' savin' the life of an inexpairienced sentry be knockin' him down off the step accident'ly an' out of the way of the muck that was flippin' iverywhere into the parodus, quiet an' vicious.

" ' Glory,' says I, as we rowled in the sewage together, ' be to the saints ! '

" ' The saints ! ' says he, scornful.

" The cap. saw us before we went down the C.T. on two stretchers. He said we done right. ' But why all the bombin' ? ' he says.

" ' We found wan of our bombs, sir, was in a dang'rous condition.'

" ' Yes ? ' says the cap. ' An' the other ? '

" ' Mebbe, sir,' says Jawn, pensive-like, ' 'twas a kind of infection.'

" ' Hum,' says the cap. ' I feared you'd be re-unitin'
the Churches.' A good off'cer. He knew all about the
two of us. Knew every man's trade in the comp'ny, an'
married or single, an' how manny in family.

" ' Gawd forbid it ! ' the two of us said in a breath.
' An' that ended the talk.' "

" And what brought the two of you here ? " I sleepily
asked.

" I was brought," Toomey said, " be a bullet woun' in
the liver, an' he be an insincaire action, a match for me
own. Wait an' I'll tell ye."

No doubt the story would have been good. But I am
only English. I am given life on terms. I have to take
sore labour's bath now and again.

" We'll go on to-morrow," I said. " A bit of shut-eye
for me now."

" Ye'll do right," he said. " Good night, an' the
blessin' of God be wid ye an' stay wid ye."

I turned over on to my right side, and snuggled in for
my sleep. The only thing I could see was the horizontal
profile in the next bed. Wasted, etherealized, abstract, the
man who had finished joy and moan had now all but
attained the remote and awful repose of a marble effigy
on a tomb in a Florentine church, seen by one who lies,
like itself, on the floor. The only thing to be heard was a
faint tap on a window above him, the delicate whipping
of some loose end of a climbing rose-tree on the glass.
And then, sudden, eruptive, winged with intention and
gusto, there came from afar the rush of a huge bass stage-
whisper : " Are ye wakin', Toomey ? "

ALL FOR PEACE AND QUIET

Deep called unto deep. "Is that Cor'pral Martin makin' night hijjus, disturbin' the ward ? "

" It is—an' good night to ye now, an' to hell wi' the Pope."

" An' the Divvil take William the Second an' William the Third, an High Dutch an' Low, an' every Martin that's in it, from Luther out, blasphemin' the whole of the day an' then late into bed, kickin' the stars of God wi' the back of your heels. So go now to your rest."

My own eyes were set on the fugitive fineness of that moribund face. It just moved, turning ever so little to right and to left as the gusts of contention blew over it. Then it settled again, the eyes always fixed on the ceiling. I thought of a water-weed on a deep pond, fluctuating minutely when gales race overhead, but soon dead-still again at its moorings.

But now the gale was abating. At each end of the ward a sequence of snorts of disdain were passing into a dying fall. Through a murmurous grumble it sank into the silent breathing of healthy infants asleep. Thus does high-handed nature interfere with the efforts of man to seek peace and ensue it.

TWO OR THREE WITNESSES

I

THANKS to our spirited Press, the eyes and the ears of the English nation are everywhere. Still, there may be, at any one time, more of these organs on one patch of ground than another. In all Connaught there were, at the time I am telling of, only four pairs of each. They were set in the heads of four London journalists, now taking their ease in their inn, at the small town of Callow, after the labours of the day.

These labours had ended at six. They had had to. The travelling Assistant Press Censor from Dublin had ordered it so : devil a word, he had said, would he take for the wire a moment later than six, and it Christmas Day in a week's time, and the girl at the post office having a long way to walk to her home in the darkness of the night.

The doyen of the four correspondents was Pellatt, who wrote for the " Day." Pellatt liked to be called the doyen. His paper, he said, was the doyen of papers. When first he had heard of that absurd six o'clock rule Pellatt had taken pity upon its author—had gone at once to see him alone and let him know what was what. The whole of the great world, as Pellatt explained, all the people who counted, read the " Day." If any parcel of a country's history did not duly impinge at the time on the consciousness of the world, that parcel of history could hardly be said to exist : at most it had only the shadowy, problematic existence of some orchid that might, or might not, have bloomed in some unexplored equatorial forest, with no one to see. Had the assistant censor no love of his country ?

At any rate, he had no vision. His mind was provincial. Weren't there four of them in it, he asked, and not Pellatt only ? Every dog had his " Day." Once give a favour and soon there'd be copy pouring into the place till God knew what hour, whereas any post-office clerk that you'd see was a human creature the like of ourselves. So six o'clock it had been on this the first day of the four press-men's labours at Callow.

It would be six to-morrow, as well. This was graver— to-morrow was going to be a big day. Thomas Curtayne, greatest of Irishmen, was to be buried in homely state at his birthplace, Kilmullen, a village some twenty miles off. Here was a sob-story chance, manifestly. The London papers were wanting to " feature " the rite : Curtayne's historic manner of giving old England a fugitive dig, now in the back and now in the stomach, during the pre-Treaty years had caught the fancy of our dispassionate nation of amateurs of the ring. Every correspondent would have to be there. And then he would have to be copious and vivid, crisp and magnanimous, right up to six.

II

Still, for the moment the four were at ease. Or three of them were. The fourth, Fane, the raw hand, was better off than at ease : his spirits were in a heavenly whirl. This was his first biggish job, and he thrilled with delight at everything that it brought ; every hurry or makeshift or jar or rebuff was adventure ; merely to sit there and hear the tremendous shop talked by the three middle-aged men, who knew all, was initiation; it gave him the raptures of swift conscious growth.

They were humane to him. Dinner, which they had allowed to be sound in the vital matters of claret and game, had been followed by bridge. "Not bad bridge either, considering——" Pellatt admitted after the last trick had fallen, while casting a disciplinary look at Fane from under his ponderous brows. Fane's failings were clearly the considerations referred to. Morris, his partner, had snubbed Fane's admissions and pleas, after each ill-starred flutter of their common cause, with repetitions of the formula "Correct mistakes and carry on," in tones of firm resignation. Fane felt that he had played like a conscript. But then he had been one—had only wanted, for his part, to look around Callow, to gorge himself with its quaint queerness. Still, everyone else's escape from an evening of pain had seemed to depend on his cutting in.

So he had played. But now, while Morris and Bute totted up, he could give himself without shame to the amusement of eyeing the room, so much more purely and deeply mid-Victorian in every colour and curve of its fittings and furniture than any room now to be easily found in fashion-following England. Odd that on rebel Ireland an older England should stamp her most durable image. Savoursome Tudor idioms were constantly piquing one's ear in the rustic speech of these wilds. One never heard them now in England. And those ghost-like "leaders" in Irish newspapers: up and down the badly-printed columns walked the spectre of good Georgian London wit; here subsisted the ironies, coquetries, breeding and ease lost to England long since, when writers grew flustered and polysyllabic and shrill and forgot how it bit when Steele was most quiet. Ah, but of course we had character still! Our oaken English uprightness was not put away, out of use, in any lavendered drawer of this still-room in Ireland. Rugged integrity—that suit, at least, we were strong in.

We must be, for Pellatt was saying it now. When Pellatt, with his short-curled grizzled hair and Jovian eye-brows and his frown, a very Elijah of Handel, positively said " 'Tis so " Fane could see it no otherwise. Pellatt was quoting some great person who had said that all our bother in Ireland had come of the desperate attempt of an honest and slow-witted race to govern a race of quick-witted scamps. " I'm not so sure," Pellatt judicially added, " about the ' slow-witted.' "

The fleeting hour was genial—no hour in which to set bounds to the great love a man bears to country and self. On the three curvilinear waistcoats before him Fane counted eight buttons undone; it gave him an intimate joy; he was in at the spot where greatness unbent. Within the three middle-aged bosoms subjacent to those straining waistcoats a similar process of kindly expansion advanced at no meaner pace. The three seasoned vessels began to confess, each after his kind, the severe morality of their professional conduct. Men are prone to do that when their bodies are deeply at ease; in that mellow condition the thought of the stern hold that you keep on yourself could move you to tears. In straight-flung words and few the handsome Bute acknowledged his creed—" Don't let your paper down, ever—that's good enough law and prophets for me." Bute bit his lip on the last word, with a manly control of emotion—Fane saw it and looked away for a moment, respecting it. Bute was a dear; Fane had already seen that; Bute's extremely kind and shy eyes, that dodged you behind shiny glasses, were always credentials enough to gain your goodwill. But now Bute the gentle concurrer with everyone else's movements, Bute the averter of all jolts and jars in your mind, the smoother of conversational ways and avoider of clashes and gloom, shone forth as more than a dear. And of course he was right: he had the key—Fane perceived it: have some one loyalty; be true to any one thing; all the rest follows.

And now Morris found tongue. Morris the rough diamond, the "sound practical journalist" of "the old school," the man who, when cross, would complain of Pellatt's "racuous" voice and of Bute's "baynal" old gobbets of Latin; Morris, who said a few minutes ago that the assistant press censor was clearly "a total stranger to the amnities of life." Yet Morris surely knew a thing or two: he must. Morris worked for no single paper. He was the great News Agency man; his messages went to all papers, morning and evening; Morris's stuff would be used by any paper that had no man of its own on the spot. Now he broke forth, as well as he could at this hour, about "the brotherhood of the craft"; journalists ought to be comrades faithful and true; news ought to be pooled; "scoops" and "beats" and any dog's trick of the kind ought to be barred; any brother-craftsman attacked by the outer world ought to be backed to the death by the rest. " I mean to say, simply, the craft! Stick together, and stick to the craft ! " As evenings wore on, Morris's speech came more and more to be composed of many " I mean to say's," each followed by some new essay in verbal expression less effective than the last. He bulged with rich meaning, rather than uttered it. Still, Fane understood. A man might have honour and grit though he did not carry his drink to perfection. Burns, no doubt, was like that of an evening. Sheridan too.

" Oh, I'm absolutely with you there," Bute was saying to Morris cordially. Pellatt was eyeing the weaker vessel with somewhat Olympian disdain, but Bute would not fail at doing the merciful thing.

" Well," said Bute, in the first pause that came, " *Cras*, I suppose, *ingens iterabimus aequor*." Bute had been a classical scholar at Corpus and seemed unashamed of the fact. " Time I crept into my narrow bed."

Morris, as soon as the door had closed upon Bute, proved how fast events had been marching within him. " Good ol' Butey ! " he gurgled. " Mean t'shay, shafety first. Minnight oil, y'know. Mean t'shay, got to pump two'r three gallonsh—fill up spare tank before starting, y'know. Mean t'shay, good ol' wise virgin, ol' Butey, 's what *I* mean."

It was cryptic to Fane. Pellatt was simpler. " He might have waited a little," said Pellatt, austerely. Bute had not left him a moment to sum the case up and put the thing in the right light. All that Pellatt did now, with so meagre an audience to awe, was just to impart, in brief, what it was that had always kept him up to the mark.

Just the one simple reflection : " Mind, he who writes for the ' Day ' writes history. He who writes history makes it."

Pellatt took care to risk no anti-climax to this high-toned aphorism. He rose, and finished his glass. And Fane was equally willing to go and sleep on all the inspiring things he had heard.

III

Lanterns were gleaming about the inn yard when the yawning waiter brought in their unloved early breakfast. Unconfirmed rumour had given out that their man would be buried at 10-30. Unconfirmed rumour ! " Good God ! " Pellatt groaned, as he dourly and doggedly ate, " is there any other country in Europe would leave you in doubt ? "

So, to make safe, and to get a look at Kilmullen first, they must start a good deal before eight. All motors and petrol had lately been commandeered by the State : nothing for it but horsed jaunting-cars ; so each of the four had chartered one of these for himself, to secure individual liberty. Just as they finished their eggs the scrape and slipping of hoofs on the rounded stones of the yard, as the first horse was led out of its stall, roused Pellatt once more : " Horsed cars for the Press ! What a country ! Savagery ! Absolute savagery ! "

" *You* needn't grouse," Morris snarled. " *You've* not got to be back in this hole at 1-30 to wire your story for evening papers."

" I've got to write a story fit to live," Pellatt growled.

Bute rushed in to create a pacific diversion. He play-fully wailed "*Quel métier ! Quel métier !* Up early, down late. Wise man, old Chaucer—' Flee from the Presse and hold by Steadfastnesse.' "

It seemed that laments were the fashion. So Fane did not let on how he inwardly chuckled and grinned with delight. The voices and flitting lights in the yard, the big day's work afoot, with Charles's Wain still bright above a darkling gable, the sweet or sour tempers of men already in play before dawn—full of delicious differences—this was life, full-flavoured : the cup had strong wine in it. Heavenly, too, was the drive out from Callow ; first through the murky pallor of failing night, growing hag-gard, and then the realist light of a cloud-covered dawn, dreary, circumstantially blank. Frost quickens fires, and darkness gives beauty to flame : that bald, aghast oncome of day, without colour or stir, seemed to brace the delighted spirit in Fane to rally its stout garrison of inward cheer for a defiance to all enveloping forces of torpor and glum-ness. On each of the three cars ahead, back to back with its driver, was seated a stolid, immobile figure, immense with greatcoats. Fane figured each of the three as a symbol of domination over investing armies of circumstance ; each was a lamp amid gloom, a fortress exultantly held, a hut with a great fire in it and all the Arctic without. These and other picturesque images visited Fane's virgin mind all the way to Kilmullen.

All Kilmullen is built round "the Square"—an open quadrangle of turf, mangy-looking with wear. Each of its sides is about a hundred yards long. The high road from

Callow passes along one of these sides. The church stands on the opposite side, near its centre. Next to the church, the priest's house is on one side, the school on the other. No post office exists. Next to the high road, at the corner nearest to Callow, a grocer's and general shop enjoys the blessing of a licence for beer and porter. Near it stands an old police barracks, now used by State troops. The rest of the village is white-washed mud cottages.

The four drivers made for the grocer's large stable yard : there a man of infinite leisure came to receive the horses, and Pellatt was first to get out the question next to all hearts. Was 10-30 really the hour ?

" Tin-thirty ! The funer'l ! " The leisurely man invigorated himself to give what was almost a hoot of derision. " Tut ! Not at all. Wan o'clock, or wan-thirty. An' early at that, with His Grace the Archbishop driving a good twenty miles from Clong to this place."

Pellatt's face could be read like a book at that moment—a black-letter book. The Irish again ! The untrustworthy Irish ! The Irish all over !

" That does it," Morris said firmly. He turned to his driver : " You needn't take that horse out. I'm going back now."

The driver pleaded the frailty of horseflesh.

" Well "—Morris relented so far—" give him a snack while I look at the church. In ten minutes we'll start." He hurried off to the church, where the man of leisure said the coffin was lying now on the tallest bier that ever you saw.

The rest followed more slowly. They would have

time to look round, in all conscience, between now and one
—" or one-thirty." Three hours wasted, out of the five
in which they might have been writing their masterpieces
at Callow, in front of a good bedroom fire apiece ! Now—
well, with luck, they might get to their writing by four
instead of at one. Fane had to admit to himself that the
cause had come a bit of a cropper. But Morris's fate was
the worst. Poor old Morris would see just nothing at all.
" Why "—his thoughts broke out into speech—" Morris's
old evening papers won't get a word about the whole thing."

" No. Not if his horse falls down dead on the road,"
Pellatt savagely grunted.

Fane stared. " Oh, of course he could say how the place
looked beforehand," said Fane.

Bute murmured, " *O sancta simplicitas !* "—goodness
knew why.

Morris was speeding out of the church as they neared it.
" So long," he said, valedictively. " Back to my post. Fool
I was ever to leave it."

" The post office, you mean ? " Pellatt jibed, with a
stony glint in his eye.

" Absolutely," said Morris. " You one-paper artists
may gad about, sniffing for atmospheres. I've got the
whole of the Press to keep posted up on the facts, and I
can't play the fool any more. So—c'rect mistakes and
carry on." He sped off towards the grocer's.

" O these agency men ! " Bute murmured mildly. At
those ribald words about one-paper artists Pellatt had
turned his back on the blasphemer. Fane marvelled. All
was not clear.

IV

When the three had taken a good look at the inside of the church they trailed out into the square and strolled on its grass and presently sat and smoked on the low tubular rail that ran round it, and Pellatt and Bute sank into melancholy, crabbed or sympathetic—and so the vacant morning dragged on. It was now that Pellatt's variety first struck Fane as being less infinite than Cleopatra's. Pellatt returned on himself. In what other country on God's earth, he acidly asked of Bute, could you not find out to-day the hour fixed for the chief public event of to-morrow? Spaniards, Arabs, Hindoos, Kaffirs were all people of system, compared with this feckless breed. The Irish were "natives" really—just natives—could never get on without some firm, upright ruler above them, someone who did not babble but did get things done and knew his own mind and told the truth and stuck to it.

Bute chimed obligingly in with the other's mood of the moment. Bute recalled the saying of some clever person : ask a Greek a question and he will give the answer most likely to profit himself; ask an Arab, and he will give the answer that runs to the fewest words; ask an Irishman, and he will give you for answer whatever might give you most pleasure. But Pellatt—first letting Bute know that all this was a chestnut—did not seem quite so sure about that disinterested desire to please. Anyhow, what an Irishman said was not evidence. " Give me, for God's sake, a country where people—at any rate, men like ourselves—are brought up to feel that lying is simply a thing that's not done."

Fane felt he was idling; he ought to be marking the words of the two senior experts in observation of life. But Fane was young: pleasure distracted him. It was delicious just to sit there and see the things that went on, the personal life of the place, all its tranquil molecular movement, not hidden but only softeningly swathed in the light mist of that dull day. A tiny ass, drawing a tiny cart laden with peat, pensively picked its way, a wavy enigmatic way, along the road bounding one side of the square, its frail shanks and delicate feet moving daintily: near its head, but not intruding word or blow on the thoughts of the ass, an aged man meandered pensively too. Hither and thither over the turf of the square three mongrel dogs prosecuted together their race's old pursuits of love, war, and the chase, checked only when some lapse into pitched battle, or else a frenzied attempt to dig a rat out of the turfy bosom of the square, brought a shouted curse from a cottage door to remind the passionate creatures of the sovereignty of law. Presently children surged out of the school-house, plunging with shrill cries of joy into the lessonless outer air as into a bath, and, this first ecstasy being spent, began to play soldiers and marched up and down, banging tin cans with extreme gravity.

Oh, it was great, the low rhythm of this rustic world, like the pulse and breath of babies asleep, awaiting great days to come and amassing strength for them! Here in the castaway farms and the tiny towns with no street lamps—here was the forest in youth, the soil storing up richness for some such unpredictable sowing of seed as impregnated Assisi and Stratford. Winter held hushed in rest-giving arms the landscape that rose beyond the low roofs; in every hedge they had passed on the road the drugged sap slept its hardest, collecting in that semi-death the power to leap up straight at the sun and break into honeysuckle and rose. And here was a place kind as winter—really the noblest of seasons, the quiet creator with forward-looking eyes.

Carried down the full stream of this reverie, Fane did not notice that Pellatt had now said all the things which the English commonly say of the Irish, when cross, except those which had been added in pure good nature by Bute, to keep Pellatt going. Fane was not disturbed at his revelling till Bute's voice made itself felt like the touch of a child's propitiatory finger on a reading elder's arm:

> " Mon dieu, mon dieu, la vie est là
>   Douce et tranquille."

Bute softly purred the much-quoted lines. All lines that Bute quoted were much-quoted lines. And then Fane saw a thing clearly—that Bute's mind, released from the call to play up to Pellatt's, instinctively sidled up to his, Fane's. To play up, to chime in, was Bute's way. But not to understand everything. Bute, like Fane, was now gazing gravely out at the life of Kilmullen; but Bute was audibly bored, and assumed that Fane was. Verlaine had been quoted with irony. " Yes, *mon dieu !* Bute continued, " *tranquille à vous faire mourir.* Imagine living in this sort of a hole—if it *could* be called life ! "

Fane did his best to be bored. He felt ashamed of his privy delight—if he knew better, if he were complete, if he had mastered his craft of observation, no doubt he would see nothing but tedium and commonplace now. He honestly tried But nature is stubborn. Behold! Kilmullen was all very good, incorrigibly good ; he could not pump enjoyment out of his mind as fast as that unseaworthy vessel allowed the stuff to pour in. " Where's Pellatt ? " he asked. He had to say something, and now he noticed that Pellatt was gone.

Bute shrugged, without any harsh innuendo. " Gone to find somebody lettered enough to be moved at the notion of having the ' Day ' in the place—should you think ? "

Fane laughed. Yes, their thunderous doyen had that little failing—quite amiable. So was Bute's gentle touch on it.

Noon struck on the village clock's cheap tinny bell. They heard the strokes out, re-filled pipes, and loafed again over the green. During the next half-hour momentous events did not abound. One of the mongrels, then sitting up on his haunches and quaffing the air with uplifted nose, suddenly turned his head round and saw with indignation a shiny black-beetle in the act of treading on his tail. The mongrel whipped round to avenge ; the beetle hung on—an event, as events go in these haunts of peace, had come into the life of the place. Nobody else was in sight, yet Fane felt that the whirl of self-baffling pursuit which ensued was converting all Kilmullen's front windows into so many convergent eyes. Like a battalion formed in hollow square, Kilmullen gazed in on its centre. Fane felt its four frontages stir as the reeds round a pond wave very slightly whenever you throw a stone into its middle ; the slow ripples presently break on the stems of the reeds and they stir and then stand still again.

Eventlessness had re-settled thoroughly down on the place by 12-29. Tired with active bliss, the three mongrels slept on the turf. The increased smoke that harbingers dinner was streaming up peacefully straight from many chimneys. Suddenly Fane felt a convulsion run through the reeds round the pond. A man, red with haste and looking important and heavily fraught, arrived on a muddy push-bicycle out of the great world without.

v

The stranger rode straight to the priest's dwelling-house ; he propped his clogged and spattered mount against the hedge and entered. During the minutes that followed Fane felt that the very smoke from the chimneys stood still to watch the priest's door. One of the mongrels came to life, rose, sniffed, and gazed expectantly towards the same quarter, his flicking nostrils closely interrogating the intervening air. Of a sudden he flopped an ear forward ; the tip of his tail wagged a note of satisfaction. The priest had come out of his door. Behind him came Pellatt.

" Downy old fox ! " Bute chuckled, admiringly.

The priest walked quickly across to the old barracks. He dropped a few words on the way to each of two men who came eagerly out from houses to intercept him. When he had spoken, each of the men fell away with an air, as Fane thought, of remitted tension ; some strain of intent expectation was off for the moment. O horror !

Pellatt was walking across to them. Fane hardly needed to know what was coming as soon as Pellatt should speak. " We're for it ! " said Pellatt. " It's put off again ! "

" My God ! Till what hour ? " said Bute.

" Three o'clock. The archbishop's shay has gone phut on the road. A crazy horsed landau, in this year of grace ! The worship of saints is modern beside it. Good God ! What a Church ! "

" You've made friends with the padre ? " said Bute un-reproachfully.

Pellatt looked at him sharply. " I felt he would like," Pellatt said, " to know that the ' Day ' would have a man here." This may have seemed to leave the two others cold. Pellatt added : " Besides, I thought we might all be the better for knowing the order of the service. Do you people want it ? "

The two took down a few notes that he gave from his book. " I really don't think that it would have done," Pellatt explained, " to have us all trooping in. He isn't the easiest man to get on with. It takes the utmost tact to get anything from him at all."

Bute speculated aloud, in his sociable way. " Three o'clock. Finish 3-30. Beat the horse all the way back to his stable, to get in by six. Means—not a moment to write in."

Pellatt looked at him hard. " The time has been," Pellatt sturdily said, " when a man could write on a car "

Bute looked at Pellatt as hard as such gentle eyes could. "Nothing like pressure," said Bute, " for squeezing the best work out of a man."

Their gallantry rendered Fane almost glad that the blow had fallen. Here was adventure made yet more adventurous; faced, too, along with stout comrades. What more could you want?

The tinny tinkle of the clock had just tolled one; the priest had re-immersed himself in his house; dinner-time had wholly depopulated the green; even the dogs had legged off to see what their several quartermasters could do. "Anyhow, let us lunch," said Pellatt, still in that tone of robust fortitude.

There was no real inn to give meals. But life could just be sustained at the grocer's on biscuits and stout. The shop had a cavernous further end, dim, smoky, and malty. Here, on two chairs and an up-ended barrel, the three ate and drank and two of them talked of the stern joys of duty done against odds, and the sunshine filling the mind that will do the dirty on nobody. Pipes and a further treatment of these topics followed the soldierly meal. Now and again one or another would step out to see that no novel turn of affairs was making history in the square.

At 1-50 Pellatt went out in his turn. At 1-55 he had not returned. Bute looked at his watch and jumped up, looking appalled. "My God!" he said, "has he made a bolt for his stable?" He ran out. Half-way to the door he turned round and said compunctiously: "Sorry, but if Pellatt goes I must go. It's a bad world. Look out for yourself." And then he was gone.

Fane, seated aloft on his up-ended barrel, was not much perturbed. Still, he marvelled. First Morris had fled from the diamond mine, empty-handed. Then Bute was in terror lest Pellatt should rob himself too. If he did, it seemed Bute would feel bound in honour to add the wreck of his own day's job to those of the two others. Really, these points of professional honour were rather fine-drawn. Would he, Fane, be expected also to post back to Callow and wire to tell his paper that he had not stayed for the show and so could say nothing about it? A little thick, that. But Oh! good! here was Bute coming back.

The first consternation was gone from Bute's face. It was a worried face, though. "He's not gone," said Bute. "His car's there. No doubt he'll turn up at the time, but I *do* wish he wouldn't fade into the landscape so freely." Fane had never seen Bute so critical.

"Oh, he'll be buying a dog," Fane jocosely suggested.

"It won't be a pup," said Bute, almost grimly. He sat down again and relaxed, or seemed to be trying to. Business permitting, the grocer would drift aft into the cavernous snuggery, drink a gift glass of his own merchandise, and fraternise warily with the alien. Borne by customers, news from the great world without would trickle at intervals into the shop, some of it overheard by Fane and his friend, some of it carried back to their lair by the grocer upon his next visitation. At 2-10 someone big and proud with poignant tidings came to buy starch and to tell the terrible shock it had given His Grace when the landau foundered, and he an old, aged man: but His Grace was the man with the courage, and he after fainting and shaken

FIERY PARTICLES

almost to bits ; nothing would do him but get a new axle
at Strones and then on to Kilmullen before the darkness of
night.

In truth it was darkening already. Midwinter days
droop young if a monstrous wainscot of massy black cloud
in the west be making a false horizon half-way up the sky.
A cellarly gloom had begun to invest the snuggery, never
well lit. Voices heard in it now were acquiring, for Fane,
the romantic tone of all voices lifted in obscurity. Two-
thirty arrived, and an elderly woman dead to the world,
a true priest's housekeeper, came to buy candles. Of this
errand it seemed she was no whit enamoured. In sombre
tones she explained its occasion to the grocer Someone
unnamed had invaded the house of the priest, " Aye, like
a wolf on the fold or a pack of locusts itself, taking all, and
then ringing the bell for more to destroy." She etched—
like other etchers, not without acid—a portrait of this
Tamburlaine of our days. " A brindle-haired man, and not
a line on his face but it was the depth of a ditch, and a black
scowl upon him would give you a right to engage he's a
judge." Fane, in the semi-darkness, saw the profile of
Bute go sharply round to attend. The artist's indignation
was waxing as she went on with her picture. " He to be
sitting beyond in the parlour, poking the fire as bold as
how-are-you, and writing the Divvil knows what on a
packet of telegraph forms he'll have stolen, I seriously
think, from Callow post office. And nothing to do him
at last but complain of the light of the beautiful lamp that
we have and ask for a candle besides from his rev'rence,
that right soon would be begging his bread with resisting

194

not evil if I wasn't there." She snorted amain as she walked away with the auxiliary illuminants so reluctantly sought.

It was growing too dim for Fane to see Bute's expression. And yet Fane knew what it was. The cock of the head, the strain of the neck may tell a good deal : we are all cats, more or less ; we can tell in the dark. Yes, Pellatt was living well on the country, and Bute's gentler spirit was taking off its hat to that robust and exacting domination. Pellatt was writing. Why, of course, he was describing the long wait, the atmosphere of Kilmullen, the lovely and wonderful things in which Fane had only wallowed or basked otiosely, all the fitting prelude to the dramatic poignancy of the coming event. Fool, fool that Fane had been, not to work too while yet it was light ! Could he start still ? Too late—it was two-forty now. "Shall we stroll round ?" said Bute, with some anxiety in his voice, but not consternation. Good old Bute, he was the man to be with ; they were the two wise men of legend, at sea in a bowl, but the senior sage was not down-hearted—only wary.

In the square the premature twilight was bringing a few lamp-lit windows into distinction. To one of these lighted panes, an uncurtained one in the house of the priest, the two men were drawn, moth-like. As they gazed, the light was strengthened within : an unshaded candle, borne in some unseen hand, crossed the square of yellow radiance and came to rest close to the head of one seated. He looked up. Yes, Pellatt, of course, and his face not genially grateful, so far as they saw. Bute chuckled ;

" He thinks they ought to have put a shade on the candle. Hullo ! "

A faint stir in another part of the square had diverted their eyes. The cyclist, visibly armed with the prestige of one on whom much hung, was just setting off up the main road. Oh, of course, to sight the archbishop's carriage and bring back the tip for the priests' humble procession to turn out and meet the great man at the skirts of Kilmullen.

The tin-voiced timepiece of the place never spoke without first clearing its throat: four or five minutes before striking an hour it rumbled or creaked and got ready. This hum-ing and ha-ing had scarcely begun when the cyclist scorched back to the square, and the priest with his decent train of assistants set forth on foot for the rencounter.

VI

Now that he did come, God's viceroy on this patch of the earth was as punctual as a secular king. The voice of tin had not been thrice uplifted before the ancient prelate rose in the halted landau and shook out his travel-creased robes. But your men of eld cannot be rushed, least of all when eld has been pretty well shaken as well as wearied and cramped with a long, chilly drive. Foot after foot must be lowered, painful inch by inch, to the ground, with stalwart aid from a domestic chaplain. And entries of state, to court or hamlet, have to be decently made. It was 3-8 before the first homage was paid and the first blessing given. Only at 3-14 did the churchward procession set out.

TWO OR THREE WITNESSES

As all the eyes and ears of England marched abreast among the lay details of this procession Fane noticed Pellatt's eyes fixed on a watch at his wrist, to the apparent neglect of all other objects of interest, sacred or profane. Pellatt was like a man timing a race. Fane glanced at Bute, and found Bute's gaze fastened on Pellatt's face as fixedly as Pellatt's on the watch. But Fane could not look at them long. There was too much to see.

At 3-20 the head of the little procession, marching in column of fours, arrived at the church. To file through the door they had to form two deep: Fane and Bute were now abreast; Pellatt had fallen back into the next file. Another two minutes, and Fane and Bute were kneeling side by side in a pew, Bute next to the gangway. During the next three minutes Fane knew, with a different order of certitude from most of his cognitions of other people, that Bute, as well as he, had lost hold upon everything else, flooded out by the inrushing sense of living fellowship in the presence of the strange event of death. It was unprofessional. Still, it does happen.

The first full force of this may have passed as they rose to their feet with the rest. And then, through the door that lay open behind them, there came, clear in the deep outer stillness of the windless afternoon in the dispeopled square, the sounds of a cart put sharply in motion—the whipped horse's scraping clutch at the ground, and the brutal scrunch of wheels on rough cobbles. In the seconds that followed the beat of hoofs grew rhythmic and swiftly diminished; and Bute, coming back to the world and glancing quickly about him, whispered to Fane, " Stolen

197

away, by the Lord—the old fox!" And then, in another
second, "Young un, I'm going. Ain't you?"

The first tremendous hum of some droneful chant was
just rolling out cumulus clouds of mystical awe from the
altar, blent with the sensuous drug of the rolling incense.
"Gosh, no!" Fane whispered back. "We'd miss the
whole thing."

"*Magnifique*," he half heard Bute protesting, and then
"*n'est pas la guerre, n'est pas la guerre*, God help you!"
And then he was aware of a vacant place at his side and of
reverent slow steps suddenly breaking into a run on the
flags outside the open church door.

But a greater event was to dwarf even this. The fugitive
steps had not long passed out of hearing, the sounds of a
second car at some distance starting and driving away had
only just passed, when, with a shock and a sudden effect of
raw edges, the service broke off: first the chanting voices,
then the fortifying instrument—everything. Voices rose
which did not seem to belong to that place. One, collected
and clear, asked aloud, "Is there anny doctor here?"
And then, "Is there anny trained nurse?"; and a sturdy
young woman who looked like business slipped out of a
pew and ran up straight towards the altar. Other voices,
escapes for helpless emotion, squealed or bleated lamenta-
tions and random futilities: "Hold him up, now, the
way he can breathe"; "Not at all. Lay His Grace on his
back till you get the wine to him"; "The Lord God help
us all—is he dead?"

Fane caught a glimpse of a limp bundle of archi-
episcopal vestments carried out to a door at one side of the

chancel. The nurse followed, to minister. Then the dismayed congregation just waited, minute by minute, the scene a tragic counterpart to that interlude, comically empty, which comes while bridal parties sign in the vestry. Fane looked at his watch when the blank wait began. It was 3·34. He ought to be on the road now, to make sure of getting his news on the wire. In ten minutes more the chance would be small. In twenty there would be pretty well none. Yet here was news in the making—an archbishop dead, it might be, in trying to bury the great layman. No, he must wait at all costs, see the thing through, and then do what he could. The minutes ebbed fast. Twenty had gone when the parish priest came out and hushed the general hum with an uplifted hand. Thank God, he said, His Grace was in no danger at all. He had fainted, and what wonder at it ? But now he was conscious ; he would be perfectly well in the morning ; so now let them all go quietly to their homes, and at ten o'clock in the morning the funeral would begin. Fane slipped out of the church ahead of the rest and ran to the grocer's yard.

Four o'clock struck as he ran. Night had fairly fallen ; the first thing Fane saw in the yard was the lamps of his waiting car shining out on the flicking ears of the homesick horse all ready in the shafts. " Is it possible we can be there before six ? " Fane hungrily asked as he jumped to his place.

' Before six ! " ejaculated the man, as if scouting a scandalous piece of detraction. " And it a blood horse ! " This was a heartener for Fane. And the patrician horse certainly leapt out of the yard and along the high-road like embodied desire. Indeed, the vibration caused by this first burst of speed gave a fantastically dissipated air to the hand-writing of Fane as he scrawled on a telegraph form, by the aid of a flash-lamp, the short message he wanted to send to his paper—just the news of the postponement and of its cause—his legs doing unaided, the while, the work of adhesion to the bounding and bumping vehicle under him.

Doubtless blood told while the horse was adding this minor hardship to the career of letters. But Fane's composition was only just finished before it became apparent that six furlongs rather than twenty miles was the racing distance of this particular thoroughbred. After five miles the thought of home seemed to recede from the animal's inconstant mind. Five o'clock came before the half-way village was reached. Two miles more and the beast insisted on walking for a short time on the flat.

The irony of the business stuck pins into Fane—the thought of Morris, Pellatt, and Bute, one after the other, driving back along this road with easy time to send the news, and none to send; and then of him, Fane, trailing home with all the news to himself, only too late to send it. Oh, if the poor weary beast could but do miracles! But how on earth, if any chance did get him to Callow in time, was he to give the others the tip? The " brother-hood of the craft "—last night he had thrilled to the words : it would be scrubby not to act on them now that a chance of giving, and not merely taking, had come to himself in his apprenticeship. Yet, yet—had he not thrilled too at Bute's ideal of never letting his own paper down? Was not his sole ownership of this piece of news a thing that belonged not to him but to his paper, not a thing to be squandered by him for his fine feeling's sake, in standing treat to his friends ? Really, it was confoundedly hard— Launcelot Gobbo's case, only with no means of telling which counsellor was conscience and which was the fiend. Well, he would follow both—rush his own message off first of all and then tear round to tell the three others.

Idle dream !—the spent horse was walking again, and Callow was seven miles off, and it was 5-21. He thought he would get out and run ; then he put it off for a minute, hearing a cyclist's bell ringing behind ; and then a better thought came. He hailed the cyclist, swiftly unfolded the crisis to him, and suggested a deal. The cyclist was open to that. He was a Callow man ; he was the man who had done that dispatch-riding duty to-day at Kilmullen. He had had riding enough for one day. On getting security for the bike he was charmed to drive home on the car in Fane's place, howbeit slowly. Fane mounted the bike and plunged into the darkness ahead. But it was almost 5-30 by now, and Callow was still all the seven miles off.

On any strange push-bicycle, well clogged with mud, a pace of fourteen miles an hour on a greasy road at night is harder to attain than many racers on cinder tracks might suppose. Besides, on a patch of slime at the outskirts of Callow Fane made an acrobatic side-slip, was well rolled in the mud, and had to spend four minutes in straightening the handle-bar. That made his lateness a doubly sure thing. It was a quarter past six when he jumped off at the censor's and bounded up, three steps at a time, towards the lair of that dragon upon the first floor. The censor had once or twice given a few minutes' grace. Possibly——

Fane met his fate on the landing. The censor was coming away. He had just locked up for the night. "Could you——?" said Fane, with an agonized look. He held out the pitiful scrawl, like a child's or a drunkard's, that he had produced on the car.

"No *bon*, me boy. *Rien ne va plus*," said the censor, an amiable sportsman really, whom Monte Carlo had educated for the discharge of this delicate duty. He looked at the wild scrawl in Fane's hand and then at the mud on Fane's coat. "'Dad, but you've been at the wake," he concluded with sympathy. " Left at the post be the whole of the field, and you only taking a glass. Tell me now, was the funer'l the great things they tell me ? "

Fane gaped at him. " Funeral ! Why, it's all off. Put off. Till to-morrow."

The censor gaped now. Then he shut both his mouth and his eyes as a man might pull down the blinds to be quiet and think. At the end of some ruminant seconds he asked : " And what about all the dull thud of the earth on the coffin ? Aye, and the women's tears raining into the grave ? " He seemed to be mustering reminiscences. " Aye, and one that dropped a spray of winter jasmine in, and she the very spit and moral of Kathleen-ni-Houlihan, that all the English know of now, mourning her son ?"

"To-morrow," said Fane; "To-morrow, Ten-thirty A.M."

The censor perceived. Fane was sober. " The three divvils ! " muttered the censor. " And not a man buried

at all ! Begob, their sins have discovered them." There he
paused for a moment; then asked, "Was that all you were
wishful to wire ? "

" Yes," said Fane, hope reviving.

" Mind what I say to you now," said the censor. " We
need the most stringent press censorship here. But we
don't poke our nose, any more than by way of a form, into
any gentleman's private affairs. Now, have you ever a
friend in London—a man with a private address ? You
have—well and good. Does he know your editor's private
name—and do you ? Well and good. Now, in five
minutes from this the post office will close. But it's only
two minutes from here, and you're young and you have a
good gallop in you. Now, take this "—the censor had been
taking out of his pocket a telegraph form and now handed
it over and Fane saw it was blank except that it bore at
its foot the censor's signed name and his stamp—" and give
me your word of honour you'll only write on it, to one
private address : ' Archbishop ill. Funeral postponed till
to-morrow. Tell O'Flaherty—that, or whatever your
editor's name is—at once.' Then your own name."

They shook hands on it. " Off with you now," said the
censor. Fane leapt down the stairs with his unhoped-for
treasure. An empty side-car was passing the street-door
below He gave the man five shillings to drive like the
wind to the inn, not three hundred yards off, and tell the
three English gentlemen there, from Fane, that the funeral
was off—just " Funeral off. Archbishop ill." Then he
pedalled his best to the post office door.

The male clerk of the place was already fingering fondly

the steel outer gate, ready to catch the divine sound of a clock striking the liberative half-hour. Within, the female clerk was buzzing off the tail end of some long message— Bute's or Pellatt's, perhaps, Fane reflected—before taking in Fane's. When she read Fane's she turned human and murmured, " Thank God ! Now I can be there."

" That's right," Fane humanly said, and official reserve was the more completely suspended because the male clerk had now fastened the outer defences and come in to join them ; the three were insensibly leagued together by being alone in a fastness.

The man heard the news. " Who was it at all, then," he asked, " that the archbishop buried to-day ? "

" No one," said Fane.

" And wasn't he standing at all," said the man, " with his feet in the slippery yellow clay thrown up from the open grave ? "

" And who told you that ? " Fane inquired.

" Amn't I just after putting the last of it on to the buzzer for London myself," said the man, " not ten minutes back ? "

" Aye, and I at it too," the girl intervened, " till this minute. And earth thudding dull on the coffin and all."

" Wasn't it all written out," the man almost argued, " by two that had seen it happen at twenty to four and the lurid winter dusk going on at the time ? "

"And, the queerest of all," said the girl, "was I to be called away from my dinner at twenty to two to wire to London the way the last rites were completed not long after one ? "

So at last Fane perceived. One does perceive so. Up to some point or other no evidence, even the strongest, teaches you anything. Then, of a sudden, you see even more of the truth than is proved. In that moment Fane saw all that Pellatt had done at the priest's, and that Bute had done in his bedroom last night, and that Morris had gone home to do. Each brother after his fashion, the brotherhood of the pen had been forging the news. The phrases of last night occurred to him now, framed in derisive inverted commas—" writing history," " don't let your paper down ! " How it all stank !

Like other forms of evaporation, that of a faith is chilling. The blank collapse in Fane's face may have refrigerated the clerks, for the deep reserve of their high calling seemed now to constrict them again. " That'll be one and twopence," the girl said with restored professional dryness. The way the man closed the steel outer gate after Fane had a sharply exclusive clang.

As Fane stood on the pavement and tried to pull all this new experience together and make something of it, a quick, dry, rubbing sound approached : Bute, running in slippers, took shape out of the darkness, now turbid with mist. He viewed the locked gate with urbane dismay that seemed to acknowledge its own comic side. He did not strive nor cry, as Pellatt, Fane thought, would have done. " *Dis aliter visum*," Bute said, almost lightly ; and then, " You did all that man could do for us. *Unus homo nobis cunctando restituit rem.* But Morris was out, running after vain things, when your messenger came, and Pellatt was in his hot bath. He'll be in hotter water to-morrow—that is, if——" He looked a rather wistful inquiry at Fane.

"Yes," said Fane, " I've wired the news."

"You did quite right," said Bute. "We asked for it. 'The gods are just, and of our pleasant vices Make whips to scourge us.'"

"I'm sorry," said Fane.

As they walked along towards the hotel Bute's arm slipped under Fane's. "I'll tell you," said Bute, "the way I got started on lying. Some silly yarn was going about that a wolf had appeared in the Lakes, and I was sent down by my paper to see. It was my first job; I was on trial. I toiled at the thing for two days, questioning farmers and shepherds until it was perfectly clear that the yarn was a hoax. I was just going back when my editor sent me a wire : ' Time you were seeing that wolf. Cannot afford waste like this.' Well, I had been engaged for three years and was longing to marry. I simply couldn't afford to be sacked. So I saw the wolf—with my little eye, I saw the wolf. That was the start. *Obsta principiis.* But it's devilish hard."

"I'm sorry," said Fane. He too was betrothed, and with him too the date was an affair of finance.

VIII

Dinner opened in gloom. Not till after the fish did the good claret do its kind office so far that Morris found tongue : " Stick together, boys—that's all there is to be done. We're all in it."

" I'm not," Fane avowed.

Pellatt glared at the tyro from under storm-laden brows. Morris jeered : " Saved by unpunctuality, eh ? "

" I was in time," Fane confessed. Why the deuce had not Bute told the others and saved him, Fane, the beastly job of inflicting these punctures ?

Pellatt looked up with his eyes, without raising his head, like a hanging judge asking a question that's sure to damn the accused. " You wired," he asked, " the news of this incredible postponement ? "

" Yes," said Fane, with the rails of a dock seeming to sprout up all round him.

" Good God !" said Pellatt—and it sounded like "The Lord have mercy on your soul !"

" I suppose," said Morris, with vitriolic moderation, " loyalty is like gallantry : thank God when you see it, but don't take it hard if there's none."

" Perhaps "—Pellatt austerely ground out the words— " it had never occurred to Fane to consider all that is lost—I mean, of course, national loss—if an accident of the kind that has happened to-day is allowed to shake public faith in the Press—in the value, as I may say, of the authorised minutes of the nation's business as they are kept by the ' Day.' "

Then Fane himself lost hold. " How the deuce," he exclaimed, " could I know you were all playing this rotten trick ? "

Pellatt straightened himself. " Trick ! " he ejaculated,

augustly. Oh, it was far from being their pleasantest
evening !

IX

But we are wondrously guided. On Fane's return from
the authentic interment next day he found in his room a
long telegram from his editor. It was forbearing, but
ominous. Due allowance was made for his youth. Still,
there was to be no more missing of first-class events, no
more rash acceptance of unsifted rumours of postpone-
ments Fortunately the paper had had a good report from
the News Agency man to fall back on ; otherwise the
let-down would have been a disaster. Fane could imagine
it all : first, the London evening papers spilling about the
streets Morris's tale of the burying, wired before two
o'clock ; then Fane's own wired message coming round-
about in, over the 'phone, and looking absurd ; then, lest
there be anything in it, telephone messages flying about,
to make sure, and word arriving that all was quite right
and the funeral certainly over—the " Day " man and
somebody else had sent it in full ; and then the pencilled
note of Fane's telephoned message going on to the spike
in the editor's room, dismissed as the unpromising blunder
of a beginner.

What good was it now to report the mere truth ? It
would not be printed. What paper liked giving the lie
to itself ? Not Fane's. He knew that, whatever else
he did not know. And so he took a day off work, like the
rest.

That was their last night in Callow. Fane may have

not been elate, but a deep contentment filled two of the others, and Bute never spoiled sport. Pellatt and he both received their editors' felicitations by wire—Pellatt, " Best thanks—first-rate piece of work " ; and Bute, " Bravo—the *chose vue* in perfection." All had seen Fane's paper now, and in their relief were prepared, each after his fashion, to let Fane down lightly. " For all the serious purposes of history," Pellatt impressively said in the genial last minutes before they all went to bed, " this fellow Curtayne *was* buried yesterday. Ten years from now he *will have been* buried yesterday."

" I'm not so sure," Morris put in. " There are these God-forsaken Irish papers. They'd say anything."

" Now, do you really often hear,' Pellatt impressively asked, " of historians going down to the British Museum to check the file of the ' Day ' by that of the ' Skibbereen Eagle ' ? History *must* have some sense of relative values in evidence."

" No," Bute admitted, " and yet—yet—do you know, I think there may be something in the ways of the Old Journalism after all." Fane felt he knew Bute now. The Butes of every age had been consenting to the stoning of the prophets, out of sheer niceness to the stoners. And yet—and yet all true-born Butes said now and then, while the stoning went merrily on : " Still, there may have been something in what the old buffer was saying." They were as open-minded as that.

" Fane," said Morris, holding up his last whisky before that last one of all which he always tried to carry upstairs in his hand, lest he perish of thirst in those upper wilds, " here's luck and forgiveness. Take care, though, next time. Mean t'shay, *trop de zèle* scuttles the ship. Well, we're none of us too old—mean t'shay, too young to learn. C'rect mishtakes and carry on. Goo'night."

Fane had come to know poor Morris too. But that was easier.

A TRADE REPORT ONLY

I

NO one has said what was wrong with The Garden, nor even why it was called by that name : whether because it had apples in it, and also a devil, like Eden ; or after Gethsemane and the agonies there ; or, again, from Proserpine's garden, because of the hush filling the foreground. All the air near you seemed like so much held breath, with the long rumble of far-away guns stretching out beyond it like some dreamful line of low hills in the distance of a landscape.

The rest of the Western Front has been well written up—much too well. The Garden alone—the Holy Terror, as some of the men used to call it—has not. It is under some sort of taboo. I think I know why. If you never were in the line there before the smash came and made it like everywhere else, you could not know how it would work on the nerves when it was still its own elfish self. And if you were there and did know, then you knew also that it was no good to try to tell people. They only said, " Oh, so you all had the wind up ? " We had. But who could say why ? How is a horse to say what it is that be-devils one empty place more than another ? He has to prick up his ears when he gets there. Then he starts sweating. That's all he knows, and it was the same story with us in The Garden. All I can do is to tell you, just roughly, the make of the place, the way that the few honest solids and liquids were fixed that came into it. They were the least part of it, really.

It was only an orchard, to look at ; all ancient apple-

trees, dead straight in the stem, with fat, wet grass under-
neath, a little unhealthy in colour for want of more sun.
Six feet above ground the lowest apple boughs all struck
out level, and kept so ; some beasts, gone in our time, must
have eaten every leaf that tried to grow lower. So the
under side of the boughs made a sort of flat awning or roof.
We called the layer of air between it and the ground The
Six-foot Seam, as we were mostly miners. The light in
this seam always appeared to have had something done to
it : sifted through branches, refracted, messed about somehow, it was not at all the stuff you wanted just at that time.
You see the like of it in an eclipse, when the sun gives a
queer wink at the earth round the edge of a black mask.
Very nice, too, in its place ; but the war itself was quite
enough out of the common—falling skies all over the
place, and half your dead certainties shaken.

We and the Germans were both in The Garden, and
knew it. But nobody showed. Everywhere else on the
front somebody showed up at last ; somebody fired. But
here nothing was seen or heard, ever. You found you were
whispering and walking on tip-toe, expecting you didn't
know what. Have you been in a great crypt at twilight
under a church, nothing round you but endless thin
pillars, holding up a low roof ? Suppose there's a wolf at
the far end of the crypt and you alone at the other, staring
and staring into the thick of the pillars, and wondering,
wondering—round which of the pillars will that grey nose
come rubbing ?

Why not smash up the silly old spell, you may say—let
a good yell, loose a shot, do any sane thing to break out ?

That's what I said till we got there. Our unit took over the place from the French. A French platoon sergeant, my opposite number, showed me the quarters and posts and the like, and I asked the usual question, " How's the old Boche ? "

" *Mais assez gentil,*" he pattered. That Gaul was not waiting to chat. While he showed me the bomb-store, he muttered something low, hurried, and blurred—" *Le bon Dieu Boche,*" I think it was, had created the orchard The Germans themselves were " *bons bourgeois* " enough, for all he had seen or heard of them—" Not a shot in three weeks. *Seulement* "—he grinned, half-shamefaced and half-confidential, as sergeant to sergeant—"*ne faut pas les embêter.*"

I knew all about that. French sergeants were always like that : dervishes in a fight when it came, but dead set, at all other times, on living *paisiblement,* smoking their pipes. *Paisiblement*—they love the very feel of the word in their mouths. Our men were no warrior race, but they all hugged the belief that they really were marksmen, not yet found out by the world. They would be shooting all night at clods, at tops of posts, at anything that might pass for a head. Oh, I knew. Or I thought so.

But no. Not a shot all the night. Nor on any other night either. We were just sucked into the hush of The Garden the way your voice drops in a church—when you go in at the door you become part of the system. I tried to think why. Did nobody fire just because in that place it was so easy for anybody to kill ? No trench could be dug; it would have filled in an hour with water filtering through from the full stream flanking The Garden. Sentries stood out among the fruit trees, behind little breastworks of sods, like the things you use to shoot grouse. These screens were merely a form; they would scarcely have slowed down a bullet. They were not defences, only symbols of things that were real elsewhere. Everything else in the place was on queer terms with reality; so were they.

II

Our first event was the shriek. It was absolutely detached, unrelated to anything seen or heard before or soon after, just like the sudden fall of a great tree on a windless day. At three o'clock on a late autumn morning, a calm moonless night, the depths of The Garden in front of our posts yielded a long wailing scream. I was making a round of our posts at the time, and the scream made me think of a kind of dream I had had twice or thrice; not a story dream, but a portrait dream; just a vivid rending vision of the face of some friend with a look on it that made me feel the brute I must have been to have never seen how he or she had suffered, and how little I had known or tried to know. I could not have fancied before that one yell could tell such a lot about anyone. Where it came from there must be some kind of hell going on that went beyond all the hells now in the books, like one of the stars that are still out of sight because the world has not lived long enough to give time for the first ray of light from their blaze to come through to our eyes.

I found the sentries jumpy. "What is it, sergeant?"

one of them almost demanded of me, as if I were the fellow in charge of the devils. "There's no one on earth," he said, "could live in that misery." Toomey himself, the red-headed gamekeeper out of the County Fermanagh, betrayed some perturbation. He hinted that "Them wans" were in it. "Who?" I asked. "Ach, the Good People," he said, with a trace of reluctance. Then I remembered, from old days at school, that the Greeks, too, had been careful; they called their Furies "The Well-disposed Ladies."

All the rest of the night there was not a sound but the owls. The sunless day that followed was quiet till 2-30 p.m., when the Hellhound appeared. He came trotting briskly out of the orchard, rounding stem after stem of the fruit trees, leapt our little pretence of barbed wire, and made straight for Toomey, then on guard, as any dog would. It was a young male black-and-tan. It adored Toomey till three, when he was relieved. Then it came capering around him in ecstasy, back to the big living cellar, a hundred yards in the rear. At the door it heard voices within and let down its tail, ready to plead lowliness and contrition before any tribunal less divine than Toomey.

The men, or most of them, were not obtrusively divine just then. They were out to take anything ill that might come. All the hushed days had first drawn their nerves tight, and then the scream had cut some of them. All bawled or squeaked in the cellar, to try to feel natural after the furtive business outside.

"Gawd a'mighty!" Looker shrilled at the entry of Toomey, "if Fritz ain't sold 'im a pup!"

Jeers flew from all parts of the smoky half-darkness.
" Where's licence, Toomey ? "

" Sure 'e's clean in th' 'ouse ? "

" 'Tain't no Dogs' 'Ome 'ere. Over the way ! "

Corporal Mullen, the ever-friendly, said to Toomey,
more mildly, " Wot ? Goin' soft ? "

" A daycent dog, corp," said Toomey.   " He's bruk wi'
the Kaiser. An' I'll engage he's through the distemper.
Like as not he'll be an Alsatian."   Toomey retailed these
commendations slowly, with pauses between, to let them
sink in.

" What'll you feed him ? " asked Mullen, inspecting
the points of the beast with charity.

" Feed 'im ! " Looker squealed. " Feed 'im into th'
incinerator ! "

Toomey turned on him. " Aye, an' be et be the rats ! "

" Fat lot o' talk about rats," growled Brunt, the White
Hope, the company's only prize-fighter. Tha'd think rats
were struttin' down fairway, shovin' folk off duck-board."

" Ah ! " Looker agreed. " An' roostin' up yer armpit."

" Thot's reet," said Brunt.

" I'll bet 'arf a dollar," said Looker, eyeing the Hellhound malignantly, " the 'Uns 'ave loaded 'im up with
plague fleas. Sent 'im acrorse. Wiv instructions."

Toomey protested. " Can't ye see the dog has been hit,
ye blind man ? " In fact, the immigrant kept his tail
licking expressively under his belly except when it lifted
under the sunshine of Toomey's regard.

Brunt rumbled out slow gloomy prophecies from the

gloom of his corner. "'E'll be tearin' 'imself t'bits wi' t'mange in a fortneet. Rat for breakfas', rat for dinner, rat for tea; bit o' rat las' thing at neet, 'fore 'e'll stretch down to 't.'"

" An' that's the first sinse ye've talked," Toomey conceded. " A rotten diet-sheet is ut. An' dirt! An' no kennel the time the roof'll start drippin'. A dog's life for a man, an' God knows what for a dog."

We felt the force of that. We all had dogs at home. The Hellhound perhaps felt our ruth in the air like a rise of temperature, for at this point he made a couple of revolutions on his wheel base, to get the pampas grass of his imagination comfortable about him, and then collapsed in a curve and lay at rest with his nose to the ground and two soft enigmatic gleams from his eyes raking the twilight recesses of our dwelling. For the moment he was relieved of the post of nucleus-in-chief for the vapours of fractiousness to condense upon.

He had a distinguished successor. The company sergeant-major, no less, came round about five minutes after with " word from the colonel." Some mischief, all our hearts told us at once. They were right too. The Corps had sent word—just what it would, we inwardly groaned. The Corps had sent word that G.H.Q.— Old G.H.Q.! At it again! we savagely thought. We knew what was coming. Yes G.H.Q. wanted to know what German unit was opposite to us. That meant a raid, of course. The colonel couldn't help it. Like all sane men below brigade staffs, he hated raids. But orders were orders. He did all he could. He sent word that if anyone brought in a German, dead or alive, on his own, by this time to-morrow, he, the colonel, would give him a fiver. Of course nobody could, but it was an offer, meant decently.

Darkness and gnashing of teeth, grunts and snarls of disgust, filled the cellar the moment the C.S.M. had departed. "Gawd 'elp us!" "A ride! In The Gawden!" "'Oo says Gawd made gawdens?" "Ow! Everythink in The Gawden is lovely!" "Come into The Gawden, Maud!" You see, the wit of most of us was not a weapon of precision. Looker came nearest, perhaps, to the point. "As if we 'ad a chawnce," he said, "to gow aht rattin' Germans, wiv a sack!"

"We gotten dog for't ahl reet," said Brunt. This was the only audible trace of good humour. Toomey looked at Brunt quickly.

Toomey was destined to trouble that afternoon; one thing came after another. At 3-25 I sent him and Brunt, with a clean sack apiece, to the sergeant-major's dug-out for the rations. They came back in ten minutes. As Toomey gave me his sack, I feared that I saw a thin train of mixed black and white dust trending across the powdered mortar floor to the door. Then I saw Looker, rage in his face, take a candle and follow this trail, stooping down, and once tasting the stuff on a wet finger-tip.

And then the third storm burst. "Christ!" Looker yelled. "If 'e ain't put the tea in the sack wiv a 'ole in it!"

We all knew that leak in a bottom corner of that special sack as we knew every very small thing in our life of small things—the cracked dixie-lid, the brazier's short leg, the way that Mynns had of clearing his throat, and Brunt of working his jaws before spitting. Of course, the sack was all right for loaves and the tinned stuff. But tea !—loose tea mixed with powdered sugar ! It was like loading a patent seed-sowing machine with your fortune in gold-dust. There was a general groan of " God help us ! " with extras. In this report I leave out, all along, a great many extras. Print and paper are dear.

Looker was past swearing. " Plyin' a piper-chise ! " he ejaculated with venom. " All owver Frawnce ! Wiv our grub ! "

Toomey was sorely distressed. He, deep in whose heart was lodged the darling vision of Toomey the managing head, the contriver, the " ould lad that was in ut," had bungled a job fit for babes. " Ah, then, who could be givin' his mind to the tea," he almost moaned, " an' he with a grand thought in ut ? "

At any other time and place the platoon would have settled down, purring, under those words. " A grand thought," " a great idaya "—when Toomey in happier days had owned to being in labour with one of these heirs of his invention, some uncovenanted mercy had nearly always accrued before long to his friends—a stew of young rabbits, two brace of fat pheasants, once a mighty wild goose. The tactician, we understood in a general way, had " put the comether upon " them. Now even those delicious memories were turned to gall. "Always the sime ! " Looker snarled at the fallen worker of wonders. " Always the sime ! Ye cawn't 'ave a bit o' wire sived up for pipe-cleanin' without 'e'll loan it off yer to go snarin' 'ares." Looker paused for a moment, gathering all the resources of wrath, and then he swiftly scaled the high top-gallant of ungraciousness : " 'E wiv the 'ole platoon workin' awye for 'im, pluckin' pawtridge an' snipes, the 'ole wye up from the sea ! Top end o' Frawnce is all a muck o' feathers wiv 'im ! "

All were good men ; Looker, like Toomey, a very good man. It was only their nerves that had gone, and the jolly power of gay and easy relentment after a jar. However they tried, they could not cease yapping. I went out for a drink of clean air. If you are to go on loving mankind, you must take a rest from it sometimes. As I went up the steps from the cellar the rasping jangle from below did not cease ; it only sank on my ears as I went. " Ow, give us 'Owm Rule for England, Gord's sike ! " " Sye there ain't no towds in Irelan', do they ? " " Looker, I've tould you I'm sorry, an'—— " " Garn, both on yer ! Ol' gas-projectors ! " " Begob, if ye want an eye knocked from ye then——! " I was going back, but then I heard Corporal Mullen, paternal and firm, like Neptune rebuking the winds. " Now, then, we don't none of us want to go losing our heads about nothing." No need to trouble. Mullen would see to the children.

III

I went east, into The Garden. Ungathered apples were going to loss on its trees. I stood looking at one of them for a time, and then it suddenly detached itself and fell to the ground with a little thud and a splash of squashed brown rottenness, as if my eye had plucked it. After that sound the stillness set in again : stillness of autumn, stillness of vigilant fear, and now the stillness of oncoming evening, the nun, to make it more cloistral. No silence so deep but that it can be deepened ! As minutes passed, infinitesimal whispers—I think from mere wisps of eddies, twisting round snags in the stream—began to lift into hearing. Deepening silence is only the rise int clearness, of this or that more confidential utterance.

I must have been sucking that confidence in for a good twenty minutes before I turned with a start. I had to, I did not know why. It seemed as if some sense, which I did not know I had got, told me that someone was stealing up behind me. No one there ; nothing but Arras, the vacuous city, indistinct among her motionless trees. She always seemed to be listening and frightened. It was as if the haggard creature had stirred.

I looked to my front again, rather ashamed. Was I losing hold too, I wondered, as I gazed level out into the Seam and watched the mist deepening ? Each evening that autumn, a quilt of very white mist would come out of the soaked soil of The Garden, lay itself out, flat and dense, but shallow at first, over the grass, and then deepen upward as twilight advanced, first submerging the tips of the grass and the purple snake-headed flowers ; and then thickening steadily up till the whole Six-foot Seam was packed with milky opaqueness.

Sixty yards out from our front a heron was standing, immobilized, in the stream, staring down—for a last bit of fishing no doubt. As I watched him, his long head came suddenly round and half up. He listened. He stood like that, warily, for a minute, then seemed to decide it was no place for him, hoisted himself off the ground, and winged slowly away with great flaps. I felt cold, and thought, " What a time I've been loafing round here ! " But I found it was four o'clock only. I thought I would go on and visit my sentries, the three-o'clock men who would come off duty at five. It would warm me ; and one or two of the young ones were apt to be creepy about sundown.

Schofield, the lad in one of our most advanced posts, was waist-deep in the mist when I reached him.

"Owt, boy ? " I whispered. He was a North-country man.

" Nowt, sergeant," he answered, " barrin'——— " He checked. He was one of the stout ones you couldn't trust to yell out for help if the Devil were at them.

" What's wrong ? " I asked pretty sharply.

" Nobbut t'way," he said slowly, " they deucks doan't seem t'be gettin' down to it to-neet." My eye followed his through the boughs to the pallid sky. A flight of wild-duck were whirling and counter-whirling aloft in some odd *pas d'inquiétude*. Yes ; no doubt our own ducks that had come during the war, with the herons and snipe, to live in The Garden, the untrodden marsh where, between the two lines of rifles never unloaded, no shot was ever heard and snipe were safe from all snipers. A good lad, Schofield ; he took a lot of notice of things. But what possessed the creatures ? What terror infested their quarters to-night ?

A TRADE REPORT ONLY

I looked Schofield over. He was as near to dead white as a tanned man can come—that is, a bad yellow. But he could be left. A man that keeps on taking notice of things he can see, instead of imagining ones that he can't, is a match for the terror that walketh by twilight. I stole on to our most advanced post of all. There I was not so sure of my man. He was Mynns. We called him Billy Wisdom, because he was a schoolmaster in civil life— some council school at Hoggerston. "What cheer, Billy?" I whispered. "Anything to report?"

The mist was armpit deep on him now, but the air quite clear above that; so that from three feet off I saw his head and shoulders well, and his bayonet; nothing else at all. He did not turn when I spoke, nor unfix his eyes from the point he had got them set on, in front of his post and a little below their own level. "All—quiet— and—correct—sergeant," he said, as if each word were a full load and had to be hauled by itself. I had once seen a man drop his rifle and bolt back overland from his post, to trial and execution and anything rather than that ever-lasting wait for a bayonet's point to come lunging up out of thick mist in front and a little below him, into the gullet, under the chin. Billy was near bolting-point, I could tell by more senses than one. He was losing hold on one bodily function after another, but still hanging on hard to something, some grip of the spirit that held from second to second, after muscle had mutinied and nerve was gone.

He had hardly spoken before a new torment wrung him. The whole landscape suddenly gave a quick shiver.

223

The single poplar, down the stream, just perceptibly shuddered and rustled, and then was dead still again. A bed of rushes, nearer us, swayed for an instant, and stood taut again. Absurd, you will say. And, of course, it was only a faint breath of wind, the only stir in the air all that day. But you were not there. So you cannot feel how the cursed place had tried to shake itself free of its curse, and had failed and fallen rigid again, dreeing its weird, and poor Billy with it. His hold on his tongue was what he lost now. He began to wail under his breath, " Christ, pity me ! Oh, suffering Christ, pity me ! " He was still staring hard to his front, but I had got a hand ready to grab at his belt when, from somewhere out in the mist before us, there came, short and crisp, the crack of a dead branch heavily trodden upon.

Billy was better that instant. Better an audible enemy, one with a body, one that could trample on twigs, than that vague infestation of life with impalpable sinisterness. Billy turned with a grin—ghastly enough, but a grin.

" Hold your fire," I said in his ear, " till I order." I made certain dispositions of bombs on a little shelf. Then we waited, listening, second by second. I think both our ears must have flicked like a mule's. But the marvel came in at the eye. We both saw the vision at just the same instant. It was some fifty yards from us, straight to our front. It sat on the top of the mist as though mist were ice and would bear. It was a dog, of the very same breed as the Hellhound, sitting upright like one of the beasts that support coats-of-arms ; all proper, too, as the heralds would say, with the black and tan hues as in life. The image gazed at us fixedly. How long? Say, twenty seconds. Then it about-turned without any visible use of its limbs, and receded some ten or twelve yards, still sitting up and now rhythmically rising and falling as though the mist it rode upon were undulating. Then it clean vanished. I thought it sank, as if the mist had ceased to bear. Billy thought the beast just melted into the air radially, all round, as rings made of smoke do

You know the crazy coolness, a sort of false presence of mind, that will come in and fool you a little bit further at these moments of staggering dislocation of cause and effect. One of these waves of mad rationalism broke on me now. I turned quickly round to detect the cinema lantern behind us which must have projected the dog's moving figure upon the white sheet of mist. None there, of course. Only the terrified city, still there, aghast, with held breath.

Then all my anchors gave together. I was adrift; there was nothing left certain. I thought, " What if all we are sure of be just a mistake, and our sureness about it conceit, and we no better than puppies ourselves to wonder that dogs should be taking their ease in mid-air and an empty orchard be shrieking ? " While I was drifting, I happened to notice the sleepy old grumble of guns from the rest of the front, and I envied those places. Sane, normal places; happy all who were there; only their earthworks were crumbling, not the last few certainties that we men think we have got hold of.

All this, of course, had to go on in my own mind behind a shut face. For Billy was one of the nerve specialists; he might get a V.C., or be shot in a walled yard at dawn, according to how he was handled. So I was pulling my wits together a little, to dish out some patter fit for his case—you know : the " bright, breezy, brotherly " bilge— when the next marvel came. A sound this time—a voice, too ; no shriek, not even loud, but tranquil, articulate, slow, and so distant that only the deathly stillness which gave high relief to every bubble that burst with a plop, out in the marsh, could bring the words to us at all. " Has annywan here lost a dog ? Annywan lost a good dog ? Hoond ? Goot Hoond ? Annywan lost a goot hoond ? "

You never can tell how things will take you. I swear I was right out of that hellish place for a minute or more, alive and free and back at home among the lost delights of Epsom Downs, between the races ; the dear old smelly crowd all over the course, and the merchant who carries a tray crying, " Oo'll 'ave a good cigar, gents ? Two pence ! 'Oo wants a good cigar ? Two pence ! 'Oo says a good smoke ? " And the sun shining good on all the bookies and crooks by the rails, the just and the unjust, all jolly and natural. Better than Lear's blasted heath and your mind running down !

You could see the relief settle on Mynns like oil going on to a burn on your hand. Have you seen an easy death in bed ?—the yielding sigh of peace and the sinking inwards, the weary job over ? It was like that. He breathed " That Irish swine ! " in a voice that made it a blessing. I felt the same, but more uneasily. One of my best was out there in the wide world, having God knew what truck with the enemy. Any Brass Hat that came loafing round might think, in his blinded soul, that

A TRADE REPORT ONLY

Toomey was fraternizing; whereas Toomey was dead or prisoner by now, or as good, unless delivered by some miracle of gumption surpassing all his previous practices against the brute creation. We could do nothing, could not even guess where he was in the fog. It had risen right up to the boughs; the whole Seam was packed with it, tight. No one but he who had put his head into the mouth of the tiger could pull it out now.

We listened on, with pricked ears. Voices we certainly heard; yes, more than one; but not a word clear. And voices were not what I harked for——it was for the shot that would be the finish of Toomey. I remembered during the next twenty minutes quite a lot of good points about Toomey. I found that I had never had a sulky word from him, for one. At the end of the twenty minutes the voices finally stopped. But no shot came. A prisoner, then?

The next ten minutes were bad. Towards the end of the two hours for which they lasted I could have fancied the spook symptoms were starting again. For out of the mist before us there came something that was not seen, or heard, or felt; no one sense could fasten upon it; only a mystic consciousness came of some approaching displacement of the fog. The blind, I believe, feel the same when they come near a lamp-post. Slowly this undefined source of impressions drew near, from out the uncharted spaces beyond, to the frontiers of hearing and sight, slipped across them and took form, at first as the queerest tangle of two sets of limbs, and then as Toomey, bearing on one shoulder a large corpse, already stiff, clothed in field-grey.

IV

" May I come in, sergeant ? " said Toomey, " an'
bring me sheaves wid me ? " The pride of 'cuteness shone
from his eyes like a lamp through the fog; his voice had
the urbanely affected humility of the consciously great.

" You may," said I, " if you've given nothing away."

" I have not," said he. " I'm an importer entirely. Me
exports are nil." He rounded the flank of the breastwork
and laid the body tenderly down, as a collector would
handle a Strad. " There wasn't the means of an identification about me. Me shoulder titles, me badge, me pay-book,
me small-book, me disc, an' me howl correspondence—I
left all beyant in the cellar. They'd not have got value that
tuk me." Toomey's face was all one wink. To value
himself on his courage would never enter his head. It was
the sense of the giant intellect within that filled him with
triumph.

I inspected the bulging eyes of the dead. " Did you
strangle him sitting ? " I asked.

" Not at all. Amn't I just after tradin' the dog for
him ? " Then, in the proper whisper, Toomey made his
report :

" Ye'll remember the whillabalooin' there was at meself
in the cellar. Leppin' they were, at the loss of the tea.
The end of it was that ' I'm goin' out now,' said I, ' to
speak to a man,' said I, ' about a dog,' an' I quitted the
place, an' the dog with me, knockin' his nose against every
lift of me heel. I'd a grand thought in me head, to make
them whisht thinkin' bad of me. Very near where the

228

lad Schofiel' is, I set out for Germ'ny, stoopin' low, to get all the use of the fog. Did you notus me, sergeant ? "

" Breaking the firewood ? " I said.

" Aye, I med sure that ye would. So I signalled."

Now I perceived. Toomey went on. " I knew, when I held up the dog on the palm of me hand, ye'd see where I was, an' where goin'. Then I wint on, deep into th' East. Their wire is nothin' at all; it's the very spit of our own. I halted among ut, and gev out a notus, in English an' German, keepin' well down in the fog to rejuce me losses. They didn't fire—ye'll have heard that. They sint for the man with the English. An', be the will o' God, he was the same man that belonged to the dog."

" ' Hans,' says I, courcheous but firm, ' the dog is well off where he is. Will you come to him quietly ? '

" I can't jus' give ye his words, but the sinse of them only. ' What are ye doin' at all,' he says, ' askin' a man to desert ? '

" There was serious trouble in that fellow's voice. It med me ashamed. But I wint on, an' only put double strength in me temptin's. ' Me colonel,' I told him, ' is offerin' five pounds for a prisoner. Come back with me now and ye'll have fifty francs for yourself when I get the reward. Think over ut well. Fifty francs down. There's a grand lot of spendin' in that. An' ye'll be wi' the dog.' As I offered him each injucement, I lifted th' an'mal clear of the fog for two seconds or three, to keep the man famished wid longin'. You have to be crool in a war. Each time that I lowered the dog I lep' two paces north, under the fog, to be-divvil their aim if they fired.

"'Ach, to hell wi' your francs an' your pounds,' says he in his ag'ny. 'Give me the dog or I'll shoot. I see where you are.'

"'I'm not there at all,' says I, 'an' the dog's in front of me bosom.'

"Ye'll understan', sergeant," Toomey said to me gravely, "that last was a ruse. I'd not do the like o' that to a dog, anny more than yourself.

"The poor divvil schewed in his juice for a while, very quiet. Then he out with an offer. 'Will ye take sivinty francs for the dog? It's the whole of me property. An' it only comes short be five francs of th' entire net profuts ye'd make on the fiver, an' I comin' with you.'

"'I will not,' says I, faint and low. It was tormint refusin' the cash.

"'Won't *annythin*' do ye,' says he in despair, 'but a live wan?'

"'Depinds,' says I pensively, playin' me fish. I held up the dog for a second again, to keep his sowl workin'.

"He plunged, at the sight of the creature. 'Couldn't ye do with a body?' he says very low.

"'Depinds,' says I, marvellin' was ut a human sacrifice he was for makin', the like of the Druids, to get back the dog.

"'Not fourteen hours back,' says he, 'he died on us.'

"'Was he wan of yourselves?' says I. 'A nice fool I'd look if I came shankin' back from the fair wid a bit of the wrong unit.'

"'He was,' says he, 'an' the best of us all.' An' then he went on, wid me puttin' in just a word now and then, or a glimpse of the dog, to keep him desirous and gabbin'.

230

There's no use in cheapenin' your wares. He let on how this fellow he spoke of had never joyed since they came to that place, an' gone mad at the finish wi' not gettin' his sleep without he'd be seein' Them Wans in a dream and hearin' the Banshies ; the way he bruk out at three in the morning that day, apt to cut anny in two that would offer to hold him. ' Here's out of it all,' he appeared to have said ; ' I've lived through iv'ry room in hell, how long, O Lord, how long, but it's glory an' victory now,' an' off an' away wid him West, through The Garden. ' Ye'll not have seen him at all ? ' says me friend. We hadn't notussed, I told him. ' We were right then,' says he ; ' he'll have died on the way. For he let a scream in the night that a man couldn't give an' live after. If he'd fetched up at your end,' says he, ' you'd have known, for he was as brave as a lion.'

" ' A livin' dog's better,' says I, ' than anny dead lion. It's a Jew's bargain you're makin'. Where's the deceased ? '

" ' Pass me the dog,' says he, ' an' I'll give you his route out from here to where he'll have dropped. It's his point of deparchure I stand at.'

" ' I'll come to ye there,' says I, ' an' ye'll give me his bearin's, an' when I've set eyes on me man I'll come back an' hand ye the dog, an' not sooner.'

" He was spaichless a moment. ' Come now,' says I, from me lair in the fog, ' wan of the two of us has to be trustful. I'll not let ye down.'

" ' Ye'll swear to come back ? ' says he in great anguish.

" I said, ' Tubbe sure.'

" ' Come on with ye, then,' he answered.

" I went stoopin' along to within six feet of his voice, the way ye'd swim under water, an' then I came to the surface. The clayey-white face that he had, an' the top of his body showed over a breastwork the moral of ours. An', be cripes, it was all right. The red figures were plain on his shoulder-strap—wan-eighty-six. Another breast-work the fellow to his was not thirty yards south. There was jus' the light left me to see that the sentry there was wan-eighty-six too. I'd inspicted the goods in bulk now, an' only had to see to me sample an' off home with it."

Toomey looked benedictively down on the long stiff frame with its Iron Cross ribbon and red worsted " 186." "An ould storm-throoper ! " Toomey commendingly said. " His friend gev me the line to him. Then he got anxious. ' Ye'll bury him fair ? ' he said. ' Is he a Prod-'stant ? ' says I, ' or a Cath'lic ? ' ' A good Cath'lic,' says he ; ' we're Bavarians here.' ' Good,' says I, ' I'll speak to Father Moloney meself.' ' An' ye'll come back,' says he, ' wi' the dog ? ' ' I will not,' says I, ' I shall hand him ye now. Ye're a straight man not to ha' shot me before. Besides, ye're a Cath'lic ? ' So I passed him the an'mal and off on me journey. Not the least trouble at all, findin' the body. The birds were all pointin' to ut. They hated ut. Faith, but that fellow had seen the quare things ! " Toomey looked down again at the monstrously staring eyes of his capture, bursting with agonies more fantastic, I thought, than any that stare from the bayoneted dead in a trench.

" The man wi' the dog," Toomey said, " may go the same road. His teeth are all knockin' together A match for your own, Billy." In trenches you did not pretend not to know all about one another, the best and the worst. In that screenless life friendship frankly condoled with weak nerves or an ugly face or black temper.

"Sergeant," said Toomey, "ye'll help me indent for the fiver? A smart drop of drink it'll be for the whole of the boys."

I nodded. "Bring him along," I said, "now."

"Well, God ha' mercy on his sowl," said Toomey, hoisting the load on to his back.

"And of all Christian souls, I pray God." I did not say it. Only Ophelia's echo, crossing my mind. How long would Mynns last? Till I could wangle his transfer to the divisional laundry or gaff?

I brought Toomey along to claim the fruit of his guile. We had to pass Schofield. He looked more at ease in his mind than before. I asked the routine question. "All correct, sergeant," he answered, "Deucks is coom dahn. Birds is all stretchin' dahn to it, proper."

Its own mephitic mock-peace was re-filling The Garden. But no one can paint a miasma. Anyhow, I am not trying to. This is a trade report only.

Printed in *Great Britain by* R. & R. CLARK, LIMITED, *Edinburgh.*